A GUIDE TO

SQL

Fifth Edition

Philip J. Pratt
Grand Valley State University

**COURSE
TECHNOLOGY** ™
★
THOMSON LEARNING

Australia • Canada • Mexico • Singapore • Spain • United Kingdom • United States

A Guide to SQL, Fifth Edition is published by Course Technology.

Managing Editor	Jennifer Normandin
Senior Vice President, Publisher	Kristen Duerr
Production Editor	Anne Valsangiacomo
Development Editor	Jessica Evans
Marketing Manager	Toby Shelton
Text Designer	Books By Design
Cover Designer	Efrat Reis

Disclaimer

Course Technology reserves the right to revise this publication and make changes from time to time in its content without notice.

The Web addresses in this book are subject to change from time to time as necessary without notice.

For more information, contact Course Technology, 25 Thomson Place, Cambridge, MA 02210; or find us on the World Wide Web at *www.course.com*.

For permission to use material from this text or product, contact us by
- Web: www.thomsonrights.com
- Phone: 1-800-730-2214
- Fax: 1-800-730-2215

ISBN 0-619-03382-7

Printed in Canada
1 2 3 4 5 WC 02 01 00 99

CONTENTS

PREFACE

SQL (Structured Query Language) is a popular computer language that is used by diverse groups such as home computer owners, owners of small businesses, end users in large organizations, and programmers. Although this book uses the SQL implementation in Oracle8 as a vehicle for teaching SQL, the chapter material, examples, and exercises in this book are designed to be completed using any SQL implementation.

A Guide to SQL, Fifth Edition is written for a wide range of teaching levels, including students taking introductory computer science classes to those students in advanced information systems classes. This textbook can be used for a stand-alone course on SQL or in conjunction with a database concepts textbook where students are required to learn SQL.

The chapters in this textbook should be covered in order. Students should complete the end-of-chapter exercises and the examples within the chapters for maximum learning. Because the content of Chapter 8 assumes that the reader has had instruction in or experience with at least one programming language, the instructor should determine whether students will understand the concepts. Students without a programming background will have difficulty understanding the topic of embedded SQL. Instructors can easily omit Chapter 8 from the textbook in situations where students are not comfortable with programming examples.

Distinguishing Features

Use of Examples

Each chapter contains multiple examples that use SQL to solve a problem. Following each example, students will read about the commands that are used to solve the stated problem, and then they will see the SQL commands used to arrive at the solution. For most students, learning though examples is the most effective way to master material. For this reason, instructors should encourage students to read the chapters at the computer and to input the commands shown in the figures.

Case Studies

One case study, using the Premiere Products database, is presented in all of the examples within the chapters, and also in the first set of exercises at the end of each chapter. Although the database is small in order to be manageable, the examples and exercises for the Premiere Products database simulate what a real business can accomplish using SQL commands. Using the same case study as examples within the chapter and in the end-of-chapter exercises ensures a high level of continuity to reinforce learning.

A different case study—the Henry Books database—is used in a second set of exercises at the end of each chapter. The second case study gives students a chance to venture out "on their own" without the direct guidance of examples from the text.

Question and Answer Sections

A special type of exercise, called a Q&A, is used within each chapter. These exercises force students to consider special issues and understand important questions before continuing with their study. The answer to each Q&A appears after the question. Students are encouraged to formulate their own answer before reading the answer provided in the textbook to ensure that they understand new material before proceeding.

Exercises

Each chapter concludes with two sets of exercises in which students use SQL commands to solve realistic problems using the Premiere Products and Henry Books databases. Unless indicated otherwise in the exercise, students should complete the exercises at the computer using SQL commands.

Reference Appendices

Two reference appendices are included at the end of this textbook. Appendix A is a SQL reference that describes the purpose and syntax for the major SQL commands. Students can use Appendix A to quickly identify how and when to use important commands. Appendix B includes a "How do I" reference, which lets students cross-reference the appropriate resource in Appendix A by searching for the answer to a question.

Answers to Odd-Numbered Exercises

Appendix C includes answers to the odd-numbered exercises in the book so students have a way of checking their progress while completing the end-of-chapter exercises.

Instructor Support

The Instructor's Manual to accompany *A Guide to SQL, Fifth Edition* contains detailed teaching information, answers to even-numbered end-of-chapter exercises, and test questions (and answers). Transparency masters are included for most of the figures in the text.

ORGANIZATION OF THE TEXTBOOK

Chapter 1: Introduction to Database Management

Chapter 1 introduces the concept of databases and database management systems using the Premiere Products and Henry Books databases as examples. Many Q&A exercises are provided in the chapter to ensure that students understand how to manipulate the database on paper before they begin working at the computer.

Chapter 2: An Introduction to SQL

In Chapter 2, students will learn about important concepts and terminology associated with relational databases. They will create and run SQL commands to create tables, use data types, and add rows to tables. Chapter 2 also discusses the role and use of nulls.

Chapter 3: Single-Table Queries

Chapter 3 is the first of two chapters on using SQL commands to query a database. The queries in Chapter 3 all involve single tables. Included in this chapter are discussions of simple and compound conditions; computed columns; the SQL BETWEEN, LIKE, and IN operators; using SQL functions; nesting queries; grouping data; and retrieving columns with null values.

Chapter 4: Multiple-Table Queries

Chapter 4 completes the discussion of querying a database by demonstrating queries that join more than one table. Included in this chapter are discussions of the SQL IN and EXISTS operators, nested subqueries, using aliases, joining a table to itself, SQL set operations, and the use of the ALL and ANY operators.

Chapter 5: Updating Data

In Chapter 5, students learn how to use the SQL COMMIT, ROLLBACK, UPDATE, INSERT, and DELETE commands to update table data. Students also learn how to create a new table from an existing table and how to change the structure of a table.

Chapter 6: Database Administration

Chapter 6 covers the database administration features of SQL, including the use of views; granting and revoking database privileges to users; creating, dropping, and using an index; using and obtaining information from the system catalog; and using integrity constraints to control data entry.

Chapter 7: Reports

Chapter 7 teaches students how to create basic and complex reports based on data in a table or view. Students will learn how to concatenate data, create a view for a report, change report column headings and formats, and add report titles. Students also will include totals and subtotals in a report and group data. The topics of scripts and spooling also are discussed.

Chapter 8: Embedded SQL

Chapter 8 is an optional chapter for those students who have completed at least one programming course. This chapter covers embedding SQL commands into a procedural language such as COBOL. Although COBOL is used as a vehicle to illustrate the concepts in this chapter, the material applies equally well to any language that supports embedding.

Included in this chapter are discussions of the use of embedded SQL to insert new rows and change and delete existing rows. Also included is a discussion of how to retrieve single rows using embedded SQL commands and how to use cursors to retrieve multiple rows.

Appendix A: SQL Reference

Appendix A includes a command reference for the major SQL clauses and operators. Students can use Appendix A as a quick resource when constructing commands. Each command includes a short description, a table that shows the required and options clauses and operators, and an example and its results.

Appendix B: "How do I" Reference

Appendix B provides students with an opportunity to ask a question, such as "How do I delete rows?," and to identify the appropriate section in Appendix A to use to find the answer. Appendix B is extremely valuable when students know what they want to accomplish, but can't remember the exact SQL command they need.

Appendix C: Answers to Odd-Numbered Exercises

Answers to the odd-numbered exercises in all chapters appear in this appendix so students can make sure that they are completing the exercises correctly.

General Notes To The Student

Embedded Questions

Each chapter contains special questions to help you learn important concepts. Sometimes the purpose of these questions is to ensure that you understand some crucial material before you proceed. In other cases, the questions are designed to give you the chance to consider some special concept in advance of its actual presentation. In all cases, the answer to each question appears immediately after the question. You can simply read the question and its answer, however, you will receive maximum benefit from the text if you take the time to determine the answer to the question and then check your answer against the one given in the text before you proceed with your reading.

End-of-Chapter Material

The end-of-chapter material consists of a summary and exercises. The summary briefly describes the material covered in the chapter. Scan the summary and make sure all the concepts are familiar to you. Following the summary are two sets of exercises. The first set uses the same Premiere Products database that is used in the examples in the chapter. First work the exercises in this set to make sure that you understand how to use the commands presented in the chapter. Then complete the second set of exercises, using the Henry

Books database, and apply what you learned in the chapter to a database that is not as familiar to you. (The answers to the odd-numbered exercises in both sets of exercises appear in Appendix C so you can check your work.)

Acknowledgments

I would like to acknowledge several individuals for their contributions in the preparation of this book. I appreciate the efforts of the following individuals who reviewed the manuscript and made many helpful suggestions: Danny Yakimchuk, University College of Cape Breton; Paul Leidig, Grand Valley State University; Misty Vermaat, Purdue University Calumet; Lorna Bowen St. George, Old Dominion University; and George Federman, Santa Barbara Community College.

The efforts of the following members of the staff at Course Technology, Inc. have been invaluable and have made this book possible: Jennifer Normandin, Managing Editor, and Anne Valsangiacomo, Production Editor.

It is always a pleasure to work with Jessica Evans, the Development Editor for this book. Thanks for all your efforts, Jess. As always, you're THE BEST!!

CHAPTER 1

Introduction to Database Management

OBJECTIVES

- Understand the Premiere Products database, a database for a distributor of appliances, housewares, and sporting goods called Premiere Products.

- Understand the Henry Books database, a database for a chain of bookstores called Henry Books.

■ The Premiere Products Database

The management of Premiere Products, a distributor of appliances, housewares, and sporting goods, has determined that the company's recent growth means that it is no longer feasible to maintain customer, order, and inventory data using its manual systems. By placing the data on a computer with a full-featured database management system, management will be able to ensure that the data is current and more accurate than in the present manual system. Management also will be able to produce a variety of useful reports. In addition, management wants to be able to ask questions concerning the data in the database and obtain answers to these questions easily and quickly.

In deciding what data must be stored in the database, management has determined that Premiere Products must maintain the following information about its sales representatives (sales reps), customers, and parts inventory:

1. The number, last name, first name, address, total commission, and commission rate for each sales rep

2. The customer number, last name, first name, address, current balance, and credit limit for each customer, as well as the number of the sales rep who represents the customer

3. The part number, part description, number of units on hand, item class, number of the warehouse where the item is stored, and unit price for each part in inventory

Premiere Products also must store information about orders. Figure 1.1 shows a sample order.

FIGURE 1.1   Sample order

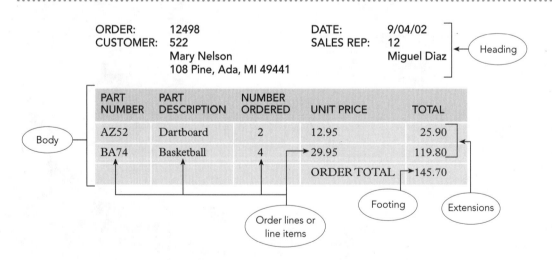

There are three components to the sample order:

1. The **heading** (top) of the order contains the order number; order date; the customer's number, name, and address; and the sales rep number and name.

2. The body of the order contains a number of **order lines**, sometimes called **line items**. Each order line contains a part number, a part description, the number of units of the ordered part, and the quoted price for the part. The order line also contains a total, usually called an **extension**, which is the product of the number ordered and the quoted price.

3. Finally, the **footing** (bottom) of the order contains the order total.

Premiere Products must also store the following items for each customer's order:

1. For each order: the order number, date the order was placed, and number of the customer who placed the order. The customer's name and address and the number of the sales rep who represents the customer are stored with the customer information. The name of the sales rep is stored with the sales rep information.

2. For each order line: the order number, part number, number of units ordered, and quoted price. Remember that the part description is stored with the information about parts. The product of the number of units ordered and the quoted price is not stored because it can be computed easily when needed.

3. The overall order total is not stored as part of the database. Instead, the total will be computed whenever an order is printed or displayed on the screen.

Figure 1.2 shows sample data for Premiere Products.

FIGURE 1.2 Sample data for Premiere Products

SALES_REP

SLSREP_NUMBER	LAST	FIRST	STREET	CITY	STATE	ZIP_CODE	TOTAL_COMMISSION	COMMISSION_RATE
03	Jones	Mary	123 Main	Grant	MI	49219	$2150.00	.05
06	Smith	William	102 Raymond	Ada	MI	49441	$4912.50	.07
12	Diaz	Miguel	419 Harper	Lansing	MI	49224	$2150.00	.05

FIGURE 1.2   Sample data for Premiere Products (continued)

CUSTOMER

CUSTOMER_ NUMBER	LAST	FIRST	STREET	CITY	STATE	ZIP_CODE	BALANCE	CREDIT_ LIMIT	SLSREP_ NUMBER
124	Adams	Sally	481 Oak	Lansing	MI	49224	$818.75	$1000	03
256	Samuels	Ann	215 Pete	Grant	MI	49219	$21.50	$1500	06
311	Charles	Don	48 College	Ira	MI	49034	$825.75	$1000	12
315	Daniels	Tom	914 Cherry	Kent	MI	48391	$770.75	$750	06
405	Williams	Al	519 Watson	Grant	MI	49219	$402.75	$1500	12
412	Adams	Sally	16 Elm	Lansing	MI	49224	$1817.50	$2000	03
522	Nelson	Mary	108 Pine	Ada	MI	49441	$98.75	$1500	12
567	Dinh	Tran	808 Ridge	Harper	MI	48421	$402.40	$750	06
587	Galvez	Mara	512 Pine	Ada	MI	49441	$114.60	$1000	06
622	Martin	Dan	419 Chip	Grant	MI	49219	$1045.75	$1000	03

ORDERS

ORDER_ NUMBER	ORDER_ DATE	CUSTOMER_ NUMBER
12489	9/02/02	124
12491	9/02/02	311
12494	9/04/02	315
12495	9/04/02	256
12498	9/05/02	522
12500	9/05/02	124
12504	9/05/02	522

ORDER_LINE

ORDER_ NUMBER	PART_ NUMBER	NUMBER_ ORDERED	QUOTED_ PRICE
12489	AX12	11	$21.95
12491	BT04	1	$149.99
12491	BZ66	1	$399.99
12494	CB03	4	$279.99
12495	CX11	2	$22.95
12498	AZ52	2	$12.95
12498	BA74	4	$24.95
12500	BT04	1	$149.99
12504	CZ81	2	$325.99

PART

PART_ NUMBER	PART_ DESCRIPTION	UNITS_ ON_HAND	ITEM_ CLASS	WAREHOUSE_ NUMBER	UNIT_ PRICE
AX12	Iron	104	HW	3	$24.95
AZ52	Dartboard	20	SG	2	$12.95
BA74	Basketball	40	SG	1	$29.95
BH22	Cornpopper	95	HW	3	$24.95
BT04	Gas Grill	11	AP	2	$149.99
BZ66	Washer	52	AP	3	$399.99
CA14	Griddle	78	HW	3	$39.99
CB03	Bike	44	SG	1	$299.99
CX11	Blender	112	HW	3	$22.95
CZ81	Treadmill	68	SG	2	$349.95

There are three sales representatives, who are identified by the numbers 03, 06, and 12. The name of sales rep number 03 is Mary Jones. Her street address is 123 Main. She lives in Grant, MI, and her zip code is 49219. Her total commission is $2,150.00, and her commission rate is 5% (.05).

Premiere Products has 10 customers who are identified with the numbers 124, 256, 311, 315, 405, 412, 522, 567, 587, and 622. The name of customer number 124 is Sally Adams. Her street address is 481 Oak. She lives in Lansing, MI, and her zip code is 49224. Her current balance is $818.75, and her credit limit is $1,000. The number 03 in the SLSREP_NUMBER column indicates that sales rep number 03 (Mary Jones) represents Sally.

Skipping down to the table named PART, there are 10 parts, which are identified by the part numbers AX12, AZ52, BA74, BH22, BT04, BZ66, CA14, CB03, CX11, and CZ81. Part AX12 is an iron and there are 104 units of this part on hand. Irons are in item class HW (housewares) and are stored in warehouse number 3. The price of an iron is $24.95. Other item classes are AP (appliances) and SG (sporting goods).

Moving back up to the table named ORDERS, there are seven orders, which are identified with the numbers 12489, 12491, 12494, 12495, 12498, 12500, and 12504. Order number 12489 was placed on September 2, 2002, by customer 124 (Sally Adams).

The table labeled ORDER_LINE might seem strange at first glance. Why do you need a separate table for the order lines? Could they be included in the ORDERS table? The answer is yes. The table labeled ORDERS could be structured as shown in Figure 1.3. Notice that this table contains the same orders as shown in Figure 1.2, with the same dates and customer numbers. In addition, each table row in Figure 1.3 contains all the order lines for a given order. Examining the fifth row, for example, you see that order 12498 has two order lines. One of these order lines is for two AZ52 parts at $12.95 each, and the other order line is for four BA74 parts at $24.95 each.

FIGURE 1.3

Sample table structure

ORDERS

ORDER_ NUMBER	ORDER_ DATE	CUSTOMER_ NUMBER	PART_ NUMBER	NUMBER_ ORDERED	QUOTED_ PRICE
12489	9/02/02	124	AX12	11	$21.95
12491	9/02/02	311	BT04	1	$149.99
			BZ66	1	$399.99
12494	9/04/02	315	CB03	4	$279.99
12495	9/04/02	256	CX11	2	$22.95
12498	9/05/02	522	AZ52	2	$12.95
			BA74	4	$24.95
12500	9/05/02	124	BT04	1	$149.99
12504	9/05/02	522	CZ81	2	$325.99

QUESTION How is the same information from Figure 1.3 represented in Figure 1.2?

ANSWER Examine the ORDER_LINE table shown in Figure 1.2 and note the sixth and seventh rows. The sixth row indicates there is an order line on order 12498 for two AZ52 parts at $12.95 each. The seventh row indicates there is an order line on order 12498 for four BA74 parts at $24.95 each. Thus, the same information that you find in Figure 1.3 is represented in Figure 1.2 in two separate rows rather than in one row.

It might seem more effective not to use two rows to represent the same information that can be represented in one row. There is a problem, however, with the arrangement shown in Figure 1.3—the table is more complicated. In Figure 1.2, there is a single entry at each location in the table. In Figure 1.3, some of the individual positions within the table contain multiple entries, thus making it difficult to track the information between columns. In the row for order number 12498, for example, it is crucial to know that the AZ52 corresponds to the 2 in the NUMBER_ORDERED column (not the 4), and that it corresponds to the $12.95 in the QUOTED_PRICE column (not the $24.95). In addition, having a more complex table means that there are practical issues to worry about, such as:

1. How much room do you allow for these multiple entries?

2. What if an order has more order lines than you have allowed room for?

3. Given a part, how do you determine which orders contain order lines for that part?

Certainly, none of these problems is unsolvable. These problems do add a level of complexity, however, that is not present in the arrangement shown in Figure 1.2. In the structure shown in Figure 1.2, there are no multiple entries to worry about, it doesn't matter how many order lines exist for any order, and finding all the orders that contain order lines for a given part is easy (just look for all order lines with the given part number in the PART_NUMBER column). In general, this simpler structure is preferable and that is why order lines appear in a separate table.

To test your understanding of the Premiere Products data, answer the following questions using the data shown in Figure 1.2.

QUESTION What are the numbers of the customers represented by Mary Jones?

ANSWER 124, 412, and 622. (Look up the number of Mary Jones in the SALES_REP table and obtain the number 03. Then find all customers in the CUSTOMER table who have the number 03 in the SLSREP_NUMBER column.)

QUESTION What is the name of the customer who placed order 12491, and what is the name of the sales rep who represents this customer?

ANSWER Don Charles is the customer, Miguel Diaz is the sales rep. (Look up the customer number in the ORDERS table and obtain the number 311. Then, find the customer in the CUSTOMER table who has number 311. Using this customer's sales rep number, which is 12, find the name of the sales rep in the SALES_REP table.)

QUESTION Which parts appear in order 12491? For each part, give the description, number ordered, and quoted price.

ANSWER. Part number: BT04, part description: Gas Grill, number ordered: 1, and quoted price: $149.99. Part number: BZ66, part description: Washer, number ordered: 1, and quoted price: $399.99. (Look up each ORDER_LINE table row in which the order number is 12491. Each of these rows contains a part number, the number ordered, and the quoted price. Use the part number to look up the corresponding part description in the PART table.)

QUESTION Why is the QUOTED_PRICE column part of the ORDER_LINE table? Can't you just take the part number and look up the price in the PART table?

ANSWER If the QUOTED_PRICE column didn't appear in the ORDER_LINE table, you would need to obtain the price for a part on an order line by looking up the price in the PART table. Although this might not be bad, it does prevent Premiere Products from charging different prices to different customers for the same part. Because Premiere Products wants the flexibility to quote and charge different prices to different customers, the QUOTED_PRICE column is included in the ORDER_LINE table. If you examine the ORDER_LINE table, you will see cases in which the quoted price matches the actual price in the PART table and cases in which it differs.

■ The Henry Books Database

Similar to the management of Premiere Products, Henry, the owner of a chain of bookstores called Henry Books, has decided it is time to computerize his operations. Like Premiere Products, he plans to store his data in a database and hopes to achieve the same benefits; that is, he wants to ensure that the data is current and accurate. He also hopes to produce several important reports. In addition, he wants to be able to ask questions concerning the data and to obtain answers to these questions easily and quickly.

In running his chain of bookstores, Henry gathers and organizes information about publishers, authors, and books. Each book has a code that uniquely identifies the book. In addition, the bookstore records the title, publisher, type of book, price, and whether the book is paperback. The bookstore also records the author or authors of the books along with the number of units of the book that are in stock in each of the four branches that make up the Henry Books chain.

Henry will store the information in a collection of tables. Figures 1.4, 1.5, and 1.6 show sample data for Henry Books.

FIGURE 1.4

· · · · · · · · · · · · · · · ·

The BRANCH,
AUTHOR, and
PUBLISHER
tables for
Henry Books

BRANCH

BRANCH_NUMBER	BRANCH_NAME	BRANCH_LOCATION	NUMBER_ EMPLOYEES
1	Henrys Downtown	16 Riverview	10
2	Henrys On The Hill	1289 Bedford	06
3	Henrys Brentwood	Brentwood Mall	15
4	Henrys Eastshore	Eastshore Mall	09

FIGURE 1.4
. . . .
The BRANCH,
AUTHOR, and
PUBLISHER
tables for
Henry Books
(continued)

PUBLISHER

PUBLISHER_CODE	PUBLISHER_NAME	PUBLISHER_CITY	PUBLISHER_STATE
AH	Arkham House Publ.	Sauk City	WI
AP	Arcade Publishing	New York	NY
AW	Addison Wesley	Reading	MA
BB	Bantam Books	New York	NY
BF	Best and Furrow	Boston	MA
JT	Jeremy P. Tarcher	Los Angeles	CA
MP	McPherson and Co.	Kingston	NY
PB	Pocket Books	New York	NY
RH	Random House	New York	NY
RZ	Rizzoli	New York	NY
SB	Schoken Books	New York	NY
SI	Signet	New York	NY
TH	Thames and Hudson	New York	NY
WN	W.W. Norton and Co.	New York	NY

AUTHOR

AUTHOR_NUMBER	AUTHOR_LAST	AUTHOR_FIRST	AUTHOR_NUMBER	AUTHOR_LAST	AUTHOR_FIRST
01	Archer	Jeffrey	13	Williams	Peter
02	Christie	Agatha	14	Kafka	Franz
03	Clarke	Arthur C.	15	Novalis	
04	Francis	Dick	16	Lovecraft	H. P.
05	Cussler	Clive	17	Paz	Octavio
06	King	Stephen	18	Camus	Albert
07	Pratt	Philip	19	Castleman	Riva
08	Adamski	Joseph	20	Zinbardo	Philip
10	Harmon	Willis	21	Gimferrer	Pere
11	Rheingold	Howard	22	Southworth	Rod
12	Owen	Barbara	23	Wray	Robert

FIGURE 1.5
....
The BOOK table
for Henry Books

BOOK

BOOK_CODE	BOOK_TITLE	PUBLISHER_CODE	BOOK_TYPE	BOOK_PRICE	PAPER-BACK
0180	Shyness	BB	PSY	7.65	Y
0189	Kane and Abel	PB	FIC	5.55	Y
0200	Stranger	BB	FIC	8.75	Y
0378	Dunwich Horror and Others	PB	HOR	19.75	N
079X	Smokescreen	PB	MYS	4.55	Y
0808	Knockdown	PB	MYS	4.75	Y
1351	Cujo	SI	HOR	6.65	Y
1382	Marcel Duchamp	PB	ART	11.25	Y
138X	Death on the Nile	BB	MYS	3.95	Y
2226	Ghost from the Grand Banks	BB	SFI	19.95	N
2281	Prints of the 20th Century	PB	ART	13.25	Y
2766	Prodigal Daughter	PB	FIC	5.45	Y
2908	Hymns to the Night	BB	POE	6.75	Y
3350	Higher Creativity	PB	PSY	9.75	Y
3743	First Among Equals	PB	FIC	3.95	Y
3906	Vortex	BB	SUS	5.45	Y
5163	Organ	SI	MUS	16.95	Y
5790	Database Systems	BF	CS	54.95	N
6128	Evil Under the Sun	PB	MYS	4.45	Y
6328	Vixen 07	BB	SUS	5.55	Y
669X	A Guide to SQL	BF	CS	23.95	Y
6908	DOS Essentials	BF	CS	20.50	Y
7405	Night Probe	BB	SUS	5.65	Y
7443	Carrie	SI	HOR	6.75	Y
7559	Risk	PB	MYS	3.95	Y
7947	dBASE Programming	BF	CS	39.90	Y
8092	Magritte	SI	ART	21.95	N
8720	Castle	BB	FIC	12.15	Y
9611	Amerika	BB	FIC	10.95	Y

WROTE

BOOK_CODE	AUTHOR_NUMBER	SEQUENCE_NUMBER
0180	20	1
0189	01	1
0200	18	1
0378	16	1
079X	04	1
0808	04	1
1351	06	1
1382	17	1
138X	02	1
2226	03	1
2281	19	1
2766	01	1
2908	15	1
3350	10	1
3350	11	2
3743	01	1
3906	05	1
5163	12	2
5163	13	1
5790	07	1
5790	08	2
6128	02	1
6328	05	1
669X	07	1
6908	22	1
7405	05	1
7443	06	1
7559	04	1
7947	07	1
7947	23	2
8092	21	1
8720	14	1
9611	14	1

INVENT

BOOK_CODE	BRANCH_NUMBER	UNITS_ON_NUMBER
0180	1	2
0189	2	2
0200	1	1
0200	2	3
079X	2	1
079X	3	2
079X	4	3
1351	1	1
1351	2	4
1351	3	2
138X	2	3
2226	1	3
2226	3	2
2226	4	1
2281	4	3
2766	3	2
2908	1	3
2908	4	1
3350	1	2
3906	2	1
3906	3	2
5163	1	1
5790	4	2
6128	2	4
6128	3	3
6328	2	2
669X	1	1
6908	2	2
7405	3	2
7559	2	2
7947	2	2
8092	3	1
8720	1	3
9611	1	2

QUESTION To check your understanding of the relationship between publishers and books, answer the following questions: Who published *Knockdown*? Which books did Signet publish?

ANSWER Pocket Books; *Cujo, Organ, Carrie*, and *Magritte*. In the row in the BOOK table (see Figure 1.5), the publisher code for *Knockdown* is PB. Examining the PUBLISHER table (see Figure 1.4), you see that PB is the code assigned to Pocket Books.

To find the books published by Signet, find its code (SI) in the PUBLISHER table. Next, find all records in the BOOK table for which the publisher code is SI. Signet published *Cujo, Organ, Carrie*, and *Magritte*.

The table named WROTE (see Figure 1.6) relates books and authors. The sequence number indicates the order in which the author of a particular text should be listed. The table named INVENT (see Figure 1.6) indicates the number of units of a particular book currently on hand at a particular branch. Row 1, for example, indicates there are two units of the book with the book code 0180 currently on hand at branch number 1.

QUESTION To check your understanding of the relationship between authors and books, answer the following questions: Who wrote *Organ*? (Make sure to list the authors in the correct order.) Which books did Jeffrey Archer write?

ANSWER Peter Williams and Barbara Owen; *Kane and Abel, Prodigal Daughter*, and *First Among Equals*. To determine who wrote *Organ*, first you examine the BOOK table to find its book code (5163). Next, look for all rows in the WROTE table in which the book code (BOOK_CODE column) is 5163. There are two such rows. In one row the author number (AUTHOR_NUMBER column) is 12, and in the other, it is 13. All that is left to do is look in the AUTHOR table to find the authors who have been assigned the numbers 12 and 13. The answer is Barbara Owen (12) and Peter Williams (13). The sequence number for author number 12 is 2, however, and the sequence number for author number 13 is 1. Thus, listing the authors in the proper order results in Peter Williams and Barbara Owen. To find the books written by Jeffrey Archer, you look up his author number in the AUTHOR_NUMBER column in the AUTHOR table and find that it is 01. Then, look for all rows in the WROTE table for which the author number is 01. There are three such rows. The corresponding book codes are 0189, 2766, and 3743. Looking up these codes in the BOOK table, you find that Jeffrey Archer wrote *Kane and Abel, Prodigal Daughter*, and *First Among Equals*.

QUESTION A customer in branch number 1 wishes to purchase *Vortex*. Is it currently in stock at branch number 1?

ANSWER　No. Looking up the code for *Vortex* in the BOOK table, you find it is 3906. To find out how many copies are in stock at branch number 1, you look for a row in the INVENT table with 3906 in the BOOK_CODE column and 1 in the BRANCH_NUMBER column. Because there is no such row, branch number 1 doesn't have any copies of Vortex.

QUESTION　You would like to obtain a copy of *Vortex* for this customer. Which other branches currently have it in stock, and how many copies does each branch have?

ANSWER　Branch number 2 has one copy, and branch number 3 has two copies. You already know that the code for *Vortex* is 3906. (If you didn't, you would look it up in the BOOK table.) To find out which branches currently have copies, look for rows in the INVENT table with 3906 in the BOOK_CODE column. There are two such rows. The first row indicates that branch number 2 currently has one copy. The second row indicates that branch number 3 currently has two copies.

SUMMARY

1. Premiere Products is an organization whose information requirements include the following:
   a. sales representatives
   b. customers
   c. orders
   d. parts
   e. order lines

2. The database for Henry Books contains the following information:
   a. branches
   b. publishers
   c. books
   d. authors
   e. inventory

EXERCISES (Premiere Products)

Answer each of the following questions using the Premiere Products data shown in Figure 1.2. No computer work is involved. In later chapters, you will use a database management system to answer questions.

1. Find the names of all the customers who have a credit limit of at least $1,500.

2. List the order numbers for orders placed by customer number 124 on September 5, 2002.

3. List the part number, part description, and on-hand value (units on hand ⋆ price) for each part in item class AP.

4. Find the number and name of each customer whose last name is Nelson.

5. How many customers have a credit limit of $1,000?

6. Find the total balance for all customers represented by sales rep number 12.

7. For each order, list the order number, order date, customer number, and customer name.

8. For each order placed on September 5, 2002, list the order number, order date, customer number, and customer name.

9. Find the number and name of each sales rep who represents any customer with a credit limit of $1,000.

10. For each order, list the order number, order date, customer number, customer name, along with the number and name of the sales rep who represents the customer.

EXERCISES (Henry Books)

Answer each of the following questions using the tables shown in Figures 1.4, 1.5, and 1.6. No computer work is involved. In later chapters, you will use a database management system to answer questions.

1. List the name of each publisher located in New York state.

2. List the name of each branch that has at least 10 employees.

3. List the code and title of each book whose type is HOR.

4. List the code and title of each book whose type is HOR and that is in paperback.

5. List the code and title of each book whose type is HOR or whose publisher code is PB.

6. List the code and title of each paperback book whose type is ART and whose price is less than $12.00.

7. Customers who are part of a special program get a 10% discount off normal

prices. For each book, list the book code, title, and discounted price. (Use the BOOK_PRICE column to calculate the discounted price.)

8. Find the code and name of each publisher for which the word "and" appears in the publisher's name.

9. List the code and title of each book with the type FIC, MYS, or ART.

10. How many books are of type MYS?

11. Find the average price for books of type HOR.

12. For each book, list the book code, title, publisher code, and publisher name.

13. For each branch, list the branch number as well as the book code, book title, and number of units on hand of each book currently in stock at the branch.

CHAPTER 2

An Introduction to SQL

OBJECTIVES

- Understand the concepts and terminology associated with relational databases

- Create and run SQL commands

- Create tables using SQL

- Identify and use data types to define the columns in SQL tables

- Understand and use nulls

- Add rows to tables

Introduction

You already might be an experienced user of a database management system (DBMS). You can find a DBMS at your school's library, at a site on the Internet, or in any other place where you retrieve data using a computer. In this chapter, you will learn about the concepts and terminology associated with the relational model for database management. Then you will learn how to create a database by describing and defining the tables and columns that make up the database. In this book, you will study a language called **SQL (Structured Query Language)**. You use SQL to manipulate data in relational databases. In SQL, you type commands to obtain the desired results.

In the mid-1970s, SQL was developed under the name SEQUEL at the IBM San Jose research facilities to be the data manipulation language for IBM's prototype relational model DBMS, System R. In 1980, the language was renamed SQL to avoid confusion with an unrelated hardware product called SEQUEL. SQL is used as the data manipulation language for IBM's current production offering in the relational DBMS arena, DB2. Most relational DBMSes use a version of SQL as a data manipulation language.

In this chapter you will learn how to assign data types for columns in the database. You will also learn about a special type of value, called a null value, and see how such values are handled during database creation. Finally, you will learn how to load a database by adding data to the tables that are created.

■ Relational Databases

A **relational database** is essentially a collection of tables like the ones you saw for Premiere Products in Chapter 1 (see Figure 2.1). A relational database is perceived by the user as being just such a collection. (The phrase "perceived by the user" simply indicates that this is how things *appear* to the user, not what the DBMS is actually doing behind the scenes.) You might wonder why this model is not called the "table" model or something similar if a database is a collection of tables. Formally, these tables are called **relations**, and this is where the model gets its name.

FIGURE 2.1   Sample data for Premiere Products

SALES_REP

SLSREP_ NUMBER	LAST	FIRST	STREET	CITY	STATE	ZIP_CODE	TOTAL_ COMMISSION	COMMISSION_ RATE
03	Jones	Mary	123 Main	Grant	MI	49219	$2150.00	.05
06	Smith	William	102 Raymond	Ada	MI	49441	$4912.50	.07
12	Diaz	Miguel	419 Harper	Lansing	MI	49224	$2150.00	.05

FIGURE 2.1 Sample data for Premiere Products (continued)

CUSTOMER

CUSTOMER_NUMBER	LAST	FIRST	STREET	CITY	STATE	ZIP_CODE	BALANCE	CREDIT_LIMIT	SLSREP_NUMBER
124	Adams	Sally	481 Oak	Lansing	MI	49224	$818.75	$1000	03
256	Samuels	Ann	215 Pete	Grant	MI	49219	$21.50	$1500	06
311	Charles	Don	48 College	Ira	MI	49034	$825.75	$1000	12
315	Daniels	Tom	914 Cherry	Kent	MI	48391	$770.75	$750	06
405	Williams	Al	519 Watson	Grant	MI	49219	$402.75	$1500	12
412	Adams	Sally	16 Elm	Lansing	MI	49224	$1817.50	$2000	03
522	Nelson	Mary	108 Pine	Ada	MI	49441	$98.75	$1500	12
567	Dinh	Tran	808 Ridge	Harper	MI	48421	$402.40	$750	06
587	Galvez	Mara	512 Pine	Ada	MI	49441	$114.60	$1000	06
622	Martin	Dan	419 Chip	Grant	MI	49219	$1045.75	$1000	03

ORDERS

ORDER_NUMBER	ORDER_DATE	CUSTOMER_NUMBER
12489	9/02/2002	124
12491	9/02/2002	311
12494	9/04/2002	315
12495	9/04/2002	256
12498	9/05/2002	522
12500	9/05/2002	124
12504	9/05/2002	522

ORDER_LINE

ORDER_NUMBER	PART_NUMBER	NUMBER_ORDERED	QUOTED_PRICE
12489	AX12	11	$21.95
12491	BT04	1	$149.99
12491	BZ66	1	$399.99
12494	CB03	4	$279.99
12495	CX11	2	$22.95
12498	AZ52	2	$12.95
12498	BA74	4	$24.95
12500	BT04	1	$149.99
12504	CZ81	2	$325.99

PART

PART_NUMBER	PART_DESCRIPTION	UNITS_ON_HAND	ITEM_CLASS	WAREHOUSE_NUMBER	UNIT_PRICE
AX12	Iron	104	HW	3	$24.95
AZ52	Dartboard	20	SG	2	$12.95
BA74	Basketball	40	SG	1	$29.95
BH22	Cornpopper	95	HW	3	$24.95
BT04	Gas Grill	11	AP	2	$149.99
BZ66	Washer	52	AP	3	$399.99
CA14	Griddle	78	HW	3	$39.99
CB03	Bike	44	SG	1	$299.99
CX11	Blender	112	HW	3	$22.95
CZ81	Treadmill	68	SG	2	$349.95

Entities, Attributes, and Relationships

There are some terms and concepts that are very important for you to know in the database environment. The terms *entity*, *attribute*, and *relationship* are fundamental when discussing databases. An **entity** is like a noun; it is a person, place, thing, or event. The entities of interest to Premiere Products, for example, are such things as customers, orders, and sales reps. The entities that are of interest to a school include students, faculty, and classes; a real estate agency is interested in clients, houses, and agents; and a used car dealer is interested in vehicles, customers, and manufacturers.

An **attribute** is a property of an entity. The term is used here exactly as it is used in everyday English. For the entity person, for example, the list of attributes might include such things as eye color and height. For Premiere Products, the attributes of interest for the entity customer are such things as first name, last name, address, city, and so on.

The final key term is *relationship*. A **relationship** is the association between entities. There is an association between customers and sales reps, for example, at Premiere Products. A sales rep is associated with all of his or her customers, and a customer is associated with his or her sales rep. Technically, you say that a sales rep is related to all of his or her customers, and a customer is related to his or her sales rep.

This particular relationship is called a **one-to-many relationship** because one sales rep is associated with many customers, but each customer is associated with only one sales rep. (In this type of relationship, the word *many* is used in a way that is different from everyday English; it might not always mean a large number. In this context, for example, the term *many* means that a sales rep may be associated with *any* number of customers. That is, one sales rep can be associated with zero, one, or more customers.)

How does a DBMS that follows the relational model handle entities, attributes of entities, and relationships between entities? Entities and attributes are fairly simple. Each entity has its own table. In the Premiere Products database, there is one table for sales reps, one table for customers, and so on. The attributes of an entity become the columns in the table. In the table for sales reps, for example, there is a column for the sales rep number, a column for the sales reps' names, and so on.

What about relationships? At Premiere Products, there is a one-to-many relationship between sales reps and customers (each sales rep is related to the *many* customers that he or she represents, and each customer is related to the *one* sales rep who represents the customer). How is this relationship implemented in a relational model database? The answer is by using common columns in two or more tables. Consider Figure 2.1 again. The SLSREP_NUMBER column of the SALES_REP table and the SLSREP_NUMBER column of the CUSTOMER table are used to implement the relationship between sales reps and customers. Given a sales rep, you can use these columns to determine all the customers that he or she represents; given a customer, you can use these columns to find the sales rep who represents the customer.

With this background, a relation is essentially a two-dimensional table. If you consider the tables shown in Figure 2.1, however, you can see that there are certain restrictions that are placed on relations. Each column should have a unique name, and entries within each column should "match" this column name. For example, if the column name is CREDIT_LIMIT, all entries in that column must be credit limits. Also, each row should be unique—if two rows are identical, the second row doesn't provide any new information. For maximum flexibility, the order of the columns and rows should be immaterial. Finally, the table should be as simple as possible. To do this you can restrict each position to a single entry by not allowing multiple entries (or **repeating groups**) in an individual location in the table (see Figure 2.2).

FIGURE 2.2   Poor table structure

ORDERS

ORDER_ NUMBER	ORDER_ DATE	CUSTOMER_ NUMBER	PART_ NUMBER	NUMBER_ ORDERED	QUOTED_ PRICE
12489	9/02/2002	124	AX12	11	$21.95
12491	9/02/2002	311	BT04	1	$149.99
			BZ66	1	$399.99
12494	9/04/2002	315	CB03	4	$279.99
12495	9/04/2002	256	CX11	2	$22.95
12498	9/05/2002	522	AZ52	2	$12.95
			BA74	4	$24.95
12500	9/05/2002	124	BT04	1	$149.99
12504	9/05/2002	522	CZ81	2	$325.99

Repeating groups

These ideas lead to the following definition:

Definition: A **relation** is a two-dimensional table in which the entries in the table are single-valued (each location in the table contains a single entry), each column has a distinct name (or attribute name), all values in a column are values of the same attribute, the order of the rows and columns is immaterial, and each row contains unique values.

From that definition, you can say that a **relational database** is a collection of relations.

Note: Rows in a table (relation) often are called **records** and columns often are called **fields**. Rows also are called **tuples** and columns are called **attributes**.

There is a commonly accepted shorthand representation of the structure of a relational database. You can write the name of the table and then, within parentheses, list all the columns (fields) in the table. For example, the following is a representation of the Premiere Products database:

```
SALES_REP (SLSREP_NUMBER, LAST, FIRST, STREET, CITY, STATE, ZIP_CODE,
    TOTAL_COMMISSION, COMMISSION_RATE)
CUSTOMER (CUSTOMER_NUMBER, LAST, FIRST, STREET, CITY, STATE, ZIP_CODE,
    BALANCE, CREDIT_LIMIT, SLSREP_NUMBER)
ORDERS (ORDER_NUMBER, ORDER_DATE, CUSTOMER_NUMBER)
ORDER_LINE (ORDER_NUMBER, PART_NUMBER, NUMBER_ORDERED, QUOTED_PRICE)
PART (PART_NUMBER, PART_DESCRIPTION, UNITS_ON_HAND, ITEM_CLASS,
    WAREHOUSE_NUMBER, UNIT_PRICE)
```

Note: In general, SQL is not case sensitive; you can type commands using uppercase or lowercase letters. There is one exception to this rule, however. When you are inserting character values into a table, you must use the correct case.

Notice that there is some duplication of names—the SLSREP_NUMBER column appears in both the SALES_REP and CUSTOMER tables. When you write SLSREP_NUMBER, how does a user of the DBMS or the DBMS itself know which SLSREP_NUMBER column you are referring to? You need a way to associate the correct table with the column name. One common approach to this problem is to write both the table name and the column name, separated by a period. Thus, the SLSREP_NUMBER column in the CUSTOMER table is written as CUSTOMER.SLSREP_NUMBER, and the SLSREP_NUMBER column in the SALES_REP table is written as SALES_REP.SLSREP_NUMBER. This technique of including the table name with the column name is known as **qualifying** the names. It is always acceptable to qualify data names, even if there is no possibility of confusion. If confusion can arise, however, it is essential to qualify the names.

The **primary key** of a table (or relation) is the column or collection of columns that uniquely identifies a given row. In the SALES_REP table, for example, the sales rep's number uniquely identifies a given row. (Sales rep 06 occurs in only one row of the table, for instance.) Thus, the SLSREP_NUMBER column is the table's primary key. You usually indicate a table's primary key by underlining the column (or collection of columns) that contains the primary key. The following is the complete shorthand representation for the Premiere Products database, where the underlined column name indicates each table's primary key:

```
SALES_REP (SLSREP_NUMBER, LAST, FIRST, STREET, CITY, STATE, ZIP_CODE,
    TOTAL_COMMISSION, COMMISSION_RATE)
CUSTOMER (CUSTOMER_NUMBER, LAST, FIRST, STREET, CITY, STATE, ZIP_CODE,
    BALANCE, CREDIT_LIMIT, SLSREP_NUMBER)
ORDERS (ORDER_NUMBER, ORDER_DATE, CUSTOMER_NUMBER)
ORDER_LINE (ORDER_NUMBER, PART_NUMBER, NUMBER_ORDERED, QUOTED_PRICE)
PART (PART_NUMBER, PART_DESCRIPTION, UNITS_ON_HAND, ITEM_CLASS,
    WAREHOUSE_NUMBER, UNIT_PRICE)
```

QUESTION   Why does the primary key for the ORDER_LINE table consist of two columns, instead of one?

ANSWER   No single column in the ORDER_LINE table uniquely identifies a given row. Two columns, ORDER_NUMBER and PART_NUMBER, are required to create a unique row.

■ Database Creation

Before you begin loading and accessing data in a table, you must describe the layout of each table to be contained in the database.

EXAMPLE 1   :   Describe the layout of the SALES_REP table to the DBMS.

The SQL command used to describe the layout of a table is CREATE TABLE. You enter the CREATE TABLE command followed by the name of the table to be created and the names and data types of the columns that comprise the table in parentheses. The **data type** indicates the type of data that the column can contain (for example, characters, numbers, or dates) as well as the maximum number of characters or digits. The rules for naming tables and columns vary slightly from one version of SQL to another. If you have any questions about naming tables or columns, consult your system's manual or your DBMS's online Help system. Some typical column naming conventions are as follows:

1. The name cannot be longer than 18 characters. (In Oracle, names can be up to 30 characters in length.)

2. The name must start with a letter.

3. The name can contain letters, numbers, and underscores (_).

4. The name cannot contain spaces.

The names used in this book should work for any SQL implementation.

The appropriate SQL command for Example 1 is shown in Figure 2.3. This SQL CREATE TABLE command, which uses the data definition features of SQL, describes a table that will be named SALES_REP. The table will contain nine columns: SLSREP_NUMBER, LAST, FIRST, STREET, CITY, STATE, ZIP_CODE, TOTAL_COMMISSION, and COMMISSION_RATE. The SLSREP_NUMBER column will contain two characters, the LAST column will contain ten characters, and the STATE column will contain two characters. The TOTAL_COMMISSION column will contain numbers only and those numbers are limited to seven digits, including two decimal places. Similarly, the COMMISSION_RATE column will contain three numbers, including two decimal places. You can visualize the SQL commands in Figure 2.3 as creating an empty table with column headings for each column name.

FIGURE 2.3 CREATE TABLE command for SALES_REP table

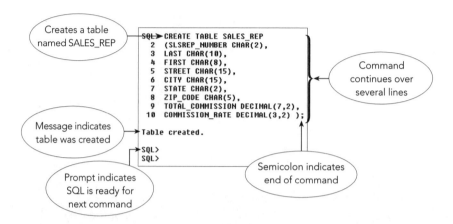

■ SQL Commands

In SQL, commands are **free format**; that is, no rule says that a particular word must begin in a particular position on the line. You could have written the previous SQL command as follows:

```
CREATE TABLE SALES_REP (SLSREP_NUMBER CHAR(2), LAST CHAR(10), FIRST CHAR(8),
STREET CHAR(15), CITY CHAR(15), STATE CHAR(2), ZIP_CODE CHAR(5),
TOTAL_COMMISSION DECIMAL(7,2), COMMISSION_RATE DECIMAL(3,2));
```

The manner in which the command is written simply makes the command more readable. This book will strive for such readability when writing SQL commands.

You can press the Enter key to end a line and then continue typing the command on the next line. You indicate the end of a command line by typing a semicolon.

Note: Ending the command with a semicolon is a requirement in Oracle, as well as in many other systems, but it is not universal.

In Oracle, the most recent command you entered is stored in the **command buffer**. You can edit the command in the buffer by using the special editing commands shown in Table 2.1.

24

Activity	Command	Abbreviation
Add text at end of current line	APPEND text	A text
Change current line	Type the line number	
Change text in current line	CHANGE /old/new	C /old/new
Delete all lines from buffer	CLEAR BUFFER	CL BUFF
Delete current line	DEL	
Edit the entire command currently in the buffer using an editor such as Notepad	EDIT	
Insert a line following current line	INPUT	I
List the command currently in the buffer	LIST	L
Run the command currently in the buffer	RUN	R

TABLE 2.1
· · · · · · · · · · · · · · · · · ·
Editing
Commands

Consider the SQL command shown in Figure 2.4.

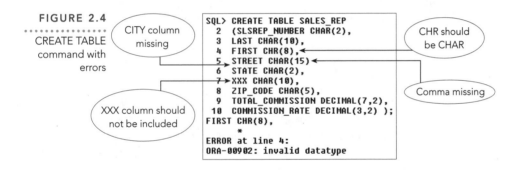

FIGURE 2.4
· · · · · · · · · · · · · · · · · ·
CREATE TABLE
command with
errors

There are several mistakes in this command. In line 4, CHAR is misspelled. Line 5 is missing a comma. The CITY column is missing, and line 7 contains errors and should be deleted.

Figure 2.5 illustrates how you can use the editing commands to make the necessary corrections. The first command, L, lists the entire SQL command in the buffer. The second command, 4, moves the current position to line 4 in the buffer. The C /CHR/CHAR command changes the misspelled CHR to CHAR. The next two commands move the current position to line 5 and add a comma to the end of the line. The 7 and DEL commands move the current position to line 7 and then delete the line. Finally the 5 and INPUT commands move the current position to line 5 and then insert a new line following the current line 5. After inserting this line, you also can insert additional lines. To indicate that you are finished adding lines, press the Enter key.

FIGURE 2.5
· · · · · · · · · · · · · · · · · · ·
Correcting a
command

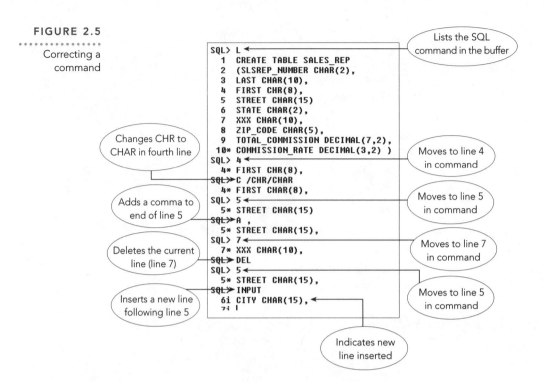

After making changes, it's a good idea to use the L command to list the modified SQL command so you can verify that your changes are correct. If you type a semicolon after a command, the command will run (or execute) immediately. To run the command you have just edited, type RUN. Typing RUN displays the command again before it is executed. If you simply want to execute the command without first displaying it, type RUN followed by a slash (/).

■ Dropping a Table

Sometimes you might find that a table in the database is no longer needed. In this case, you can use the DROP TABLE command to delete it. The command DROP TABLE is followed by the name of the table you want to delete and a semicolon. To delete the SALES_REP table, for example, you would enter the following command:

```
DROP TABLE SALES_REP;
```

Suppose in your CREATE TABLE command you inadvertently type LST instead of LAST or you type CHAR(5) instead of CHAR(15). Suppose you don't discover the error until the command is executed. In this case, you can delete the entire table using the DROP TABLE command and then create the correct table using the CREATE TABLE command. Later in this book, you will learn how to change a table's structure without having to delete the entire table.

You should note that when you drop a table, you also drop any data that you entered into the table. It's a good idea to check your CREATE TABLE commands carefully and to correct any problems before adding data.

■ Data Types

There are other data types besides DECIMAL (numeric) and CHAR (text). The actual data types vary somewhat from one implementation of SQL to another; Table 2.2 shows the commonly used data types.

TABLE 2.2

Common data types

Data Type	Description
CHAR (n)	Character string *n* characters long. Columns that contain numbers but will not be used for arithmetic operations usually are assigned a data type of CHAR. The CUSTOMER_NUMBER column, for example, is a CHAR column because the customer numbers will not be used in any calculations.
DATE	Dates in the form DD-MON-YYYY or MM/DD/YYYY. For example, May 12, 2002 could be stored as 12-MAY-2002 or as 5/12/2002. (**Note:** The specific format in which the dates are stored varies from one implementation of SQL to another.)
DECIMAL (p,q)	Decimal number *p* digits long with *q* of these being decimal places to the right of the decimal point. The data type DECIMAL(5,2) represents a number with three places to the left of the decimal point and two places to the right of the decimal point (for example, 100.00). (**Note:** The specific meaning of DECIMAL varies from one implementation of SQL to another. In some implementations, the decimal point counts as one of the places, and in other implementations it does not. Likewise, in some implementations a minus sign counts as one of the places, but in others it does not.)
INTEGER	Integers (numbers without a decimal part); the acceptable range is -2147483648 to 2147483647.
SMALLINT	Similar to INTEGER but does not occupy as much space; the acceptable range is -32768 to 32767. This data type is a better choice than INTEGER if you are certain that numbers will be within the indicated range.

■ Nulls

Occasionally, when you enter a new row into a table or modify an existing row, the values for one or more columns are unknown or unavailable. For example, you can add a customer's name and address to a table even though he or she does not have an assigned sales rep or an established credit limit. In other cases, some values might never be known; perhaps there is a customer who does not have a sales rep. In SQL you handle this problem by using a special value to represent situations in which an actual value is unknown or not applicable. This special value is called a **null data value**, or simply a **null**. When creating each column in a table, you can choose whether to allow nulls.

Q UESTION   Should a user be allowed to enter null values for the primary key?

ANSWER   No, it doesn't make sense to allow a user to enter null values for the primary key. For example, the wisdom of storing a record for a customer whose customer number is unknown is questionable at best. If you store two customer records without values in the primary key column, you will have no way to tell them apart.

Implementation of Nulls

You must have a mechanism to indicate which columns cannot contain null values. You do this by using the NOT NULL clause within the CREATE TABLE command. Those columns whose description includes NOT NULL cannot accept null values. Other columns can accept null values.

For example, suppose that the SLSREP_NUMBER, LAST, and FIRST columns in the SALES_REP table cannot accept null values, but all other columns in the SALES_REP table can. The following CREATE TABLE command would accomplish this goal:

```
CREATE TABLE SALES_REP
(SLSREP_NUMBER CHAR(2) NOT NULL,
LAST CHAR(10) NOT NULL,
FIRST CHAR(8) NOT NULL,
STREET CHAR(15),
CITY CHAR(15),
STATE CHAR(2),
ZIP_CODE CHAR(5),
TOTAL_COMMISSION DECIMAL(7,2),
COMMISSION_RATE DECIMAL(3,2) );
```

The system will reject any attempt to store a null value in either the SLSREP_NUMBER, LAST, or FIRST columns. The system will accept an attempt to store a null value in the STREET column, however, because the STREET column can accept null values.

■ Loading a Table with Data

After the tables have been created, you can load data into them. To load a table with data, you add the necessary rows to each table using the INSERT command.

The INSERT Command

When adding rows to character columns, make sure you enclose the values in single quotation marks (for example, 'Jones').

Note: You must enclose values in single quotation marks for any column whose type is character (CHAR), even if the data contains numbers. Because the ZIP_CODE column has a CHAR data type, for example, you must enclose zip codes in single quotation marks, even though they are numbers.

EXAMPLE 2 ⋮ Add sales rep 03 from Figure 2.1 to the database.

The command to add records is the INSERT command. You type INSERT INTO followed by the name of the table into which you are adding data. Then you type the VALUES command followed by the specific values to be inserted in parentheses. The command for this example and its results appear in Figure 2.6. Note that the character strings ('03', 'Jones', 'Mary', and so on) are enclosed in single quotation marks because they are values for CHAR columns.

FIGURE 2.6
·················

INSERT
command

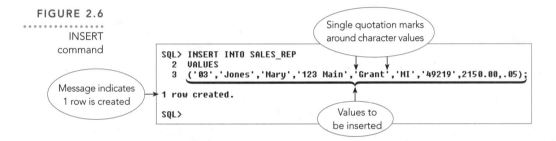

```
SQL> INSERT INTO SALES_REP
  2  VALUES
  3  ('03','Jones','Mary','123 Main','Grant','MI','49219',2150.00,.05);
1 row created.
SQL>
```

Single quotation marks around character values

Message indicates 1 row is created

Values to be inserted

Note: Make sure that you type the values in the same case as those shown in the figures to avoid problems later when retrieving data from the database.

EXAMPLE 3 ⋮ Add the second and third sales reps to the table.

You could use the INSERT command to add the rows to the table. However, an easier, faster way is to modify the previous INSERT command to add the record for the second sales rep. In Figure 2.7, the L command lists the command currently in the buffer. In this case, it is the previous INSERT command. When you type the EDIT command, SQL lets you use an editor, such as Notepad, to update the command in the buffer.

FIGURE 2.7

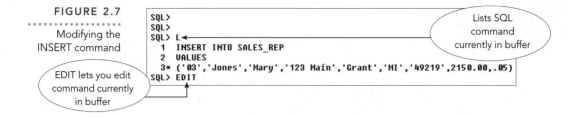

Figure 2.8 shows the command from the buffer in the Notepad program window. Figure 2.9 shows the edited command. Now you can save the changes, close the editor, and type RUN to execute the modified command. SQL will list the new command and then show you that one row was created in the table.

FIGURE 2.8   Using an editor to modify the INSERT command

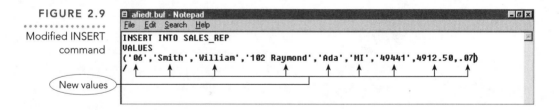

FIGURE 2.9

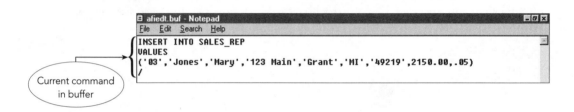

Modified INSERT
command

Figure 2.10 shows how to use the EDIT command to add the second and third records to the SALES_REP table.

FIGURE 2.10
·············
Inserting
additional rows

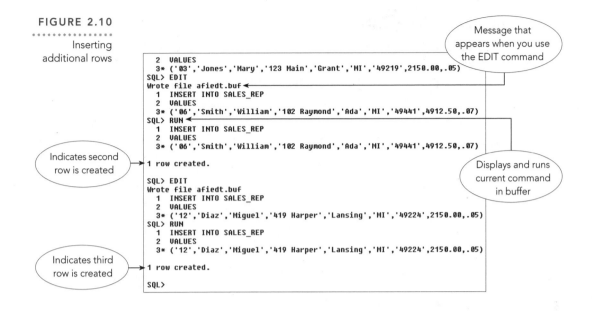

The message "Message that appears when you use the EDIT command" points to:
```
    2  VALUES
    3* ('03','Jones','Mary','123 Main','Grant','MI','49219',2150.00,.05)
SQL> EDIT
Wrote file afiedt.buf ◄
    1  INSERT INTO SALES_REP
    2  VALUES
    3* ('06','Smith','William','102 Raymond','Ada','MI','49441',4912.50,.07)
SQL> RUN ◄
    1  INSERT INTO SALES_REP
    2  VALUES
    3* ('06','Smith','William','102 Raymond','Ada','MI','49441',4912.50,.07)

1 row created.

SQL> EDIT
Wrote file afiedt.buf
    1  INSERT INTO SALES_REP
    2  VALUES
    3* ('12','Diaz','Miguel','419 Harper','Lansing','MI','49224',2150.00,.05)
SQL> RUN
    1  INSERT INTO SALES_REP
    2  VALUES
    3* ('12','Diaz','Miguel','419 Harper','Lansing','MI','49224',2150.00,.05)

1 row created.

SQL>
```

Callouts: "Indicates second row is created" points to "1 row created." (first). "Displays and runs current command in buffer" points to the RUN command area. "Indicates third row is created" points to "1 row created." (second).

The INSERT Command with Nulls

To enter a null value into a table, you use a special format of the INSERT command. In this special format, you identify the names of the columns that will accept non-null values, and then list only these non-null values after the VALUES command. See Example 4.

EXAMPLE 4 : Add sales rep 18 to the table. Her name is Elyse Martin. All columns except SLSREP_NUMBER, LAST, and FIRST are null.

In this case you do not enter a value of null; you enter only the non-null values. To do so, you must indicate precisely which values you are entering by listing the corresponding columns as shown in Figure 2.11. The command shown in the figure indicates that you are entering data in only the SLSREP_NUMBER, LAST, and FIRST columns, and that you *will not* enter a value into any other column.

FIGURE 2.11   Inserting a row containing null values
···

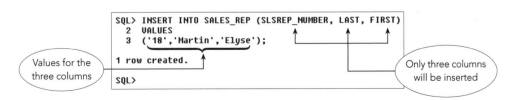

```
SQL> INSERT INTO SALES_REP (SLSREP_NUMBER, LAST, FIRST)
    2  VALUES
    3  ('18','Martin','Elyse');

1 row created.

SQL>
```

Callouts: "Values for the three columns" points to ('18','Martin','Elyse'). "Only three columns will be inserted" points to the column list.

Viewing Data in a Table

You can view data in the table to make sure that it is entered correctly by using the SQL SELECT command shown in Figure 2.12. You can use the horizontal scroll bar to see the data if it scrolls off the screen. In some implementations of SQL, extra data will wrap to a second line for improved readability.

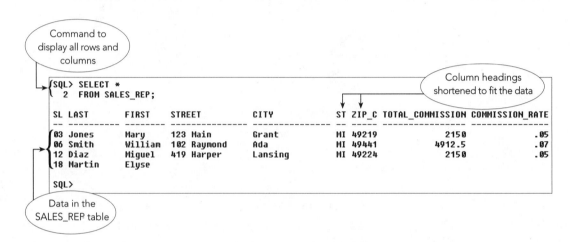

Correcting Errors in the Database

After reviewing the data in the table, you might find that you need to change the value in a column. You can use the UPDATE command shown in Figure 2.13 to correct errors. The command shown in the figure changes to Marlin the last name in the row in which the sales rep number is 18.

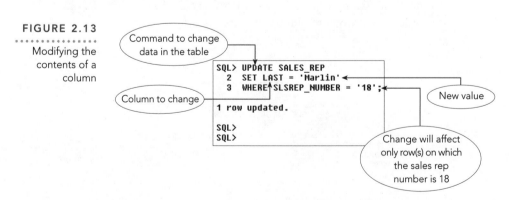

If you need to delete a record, you can use the DELETE command. The command in Figure 2.14 deletes any row on which the sales rep number is 18.

FIGURE 2.14 Deleting a row

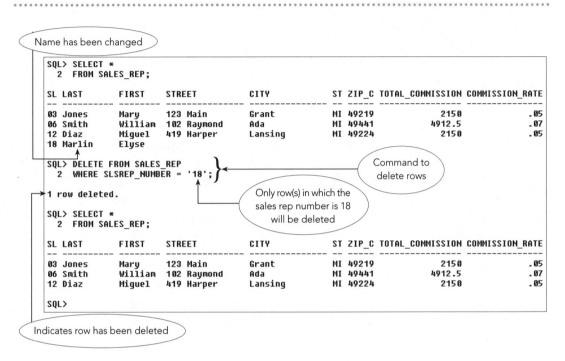

■ Additional Tables

When you are creating tables and adding rows to them, you can use an editor to create a file containing the CREATE TABLE and INSERT commands. Doing this saves you from having to type the commands over and over. To create a file in Oracle, type EDIT followed by the name of the file you want to create. Oracle assigns the file the extension .SQL automatically.

Figure 2.15 illustrates the process. The file being created is named cre_cust.SQL and it contains a CREATE TABLE command for the CUSTOMER table. Notice that two columns, LAST and FIRST, are specified as NOT NULL. Additionally, the CUSTOMER_NUMBER column is the table's primary key, indicating that the CUSTOMER_NUMBER column will be the unique identifier of rows within the table. With this column designated as the primary key, the DBMS will reject any attempt to store a customer number if that number already exists in the table. (No primary key was specified for the SALES_REP table. You will see how to add a primary key later in the book.)

FIGURE 2.15   CREATE TABLE command for CUSTOMER table

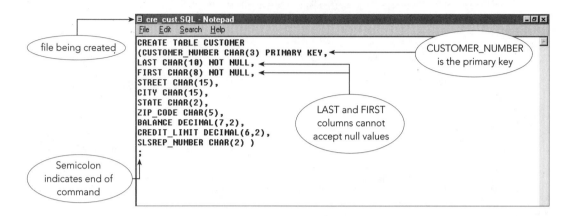

To run this command, save the file and then close the editor. Type @ (the "at" symbol) followed by the name of the file. In this case, you would type @cre_cust. (If the SQL file is stored in a folder other than the default folder for your DBMS, type the full path to the file, including the drive letter.) After you press the Enter key, SQL executes the command and creates the table.

After creating the table, you could create another file containing all the necessary INSERT commands to add the necessary records to the table. Each command must end with a semicolon. Figure 2.16 shows such a file for the CUSTOMER table.

FIGURE 2.16   INSERT commands for CUSTOMER table

```
INSERT INTO CUSTOMER
VALUES
('124','Adams','Sally','481 Oak','Lansing','MI','49224',818.75,1000,'03');
INSERT INTO CUSTOMER
VALUES
('256','Samuels','Ann','215 Pete','Grant','MI','49219',21.50,1500,'06');
INSERT INTO CUSTOMER
VALUES
('311','Charles','Don','48 College','Ira','MI','49034',825.75,1000,'12');
INSERT INTO CUSTOMER
VALUES
('315','Daniels','Tom','914 Cherry','Kent','MI','48391',770.75,750,'06');
INSERT INTO CUSTOMER
VALUES
('405','Williams','Al','519 Watson','Grant','MI','49219',402.75,1500,'12');
INSERT INTO CUSTOMER
VALUES
('412','Adams','Sally','16 Elm','Lansing','MI','49224',1817.50,2000,'03');
INSERT INTO CUSTOMER
VALUES
('522','Nelson','Mary','108 Pine','Ada','MI','49441',98.75,1500,'12');
INSERT INTO CUSTOMER
VALUES
('567','Dinh','Tran','808 Ridge','Harper','MI','48421',402.40,750,'06');
INSERT INTO CUSTOMER
VALUES
('587','Galvez','Mara','512 Pine','Ada','MI','49441',114.60,1000,'06');
INSERT INTO CUSTOMER
VALUES
('622','Martin','Dan','419 Chip','Grant','MI','49219',1045.75,1000,'03');
```

Data for first row in the table

Data for second row in the table

Each command ends with a semicolon

Figures 2.17 through 2.22 show files containing the necessary CREATE TABLE and INSERT commands for the other tables in the Premiere Products database. Figure 2.17 contains a CREATE TABLE command for the ORDERS table, and Figure 2.18 contains INSERT commands for the rows in the ORDERS table. Notice the way dates are entered.

FIGURE 2.17
. . . .
CREATE TABLE
command for
ORDERS table

```
cre_ord.SQL - Notepad
File  Edit  Search  Help
CREATE TABLE ORDERS
(ORDER_NUMBER CHAR(5) PRIMARY KEY, ◄──────
ORDER_DATE DATE,
CUSTOMER_NUMBER CHAR(3) )
;
```

ORDER_NUMBER
is the primary key

FIGURE 2.18   INSERT commands for ORDERS table

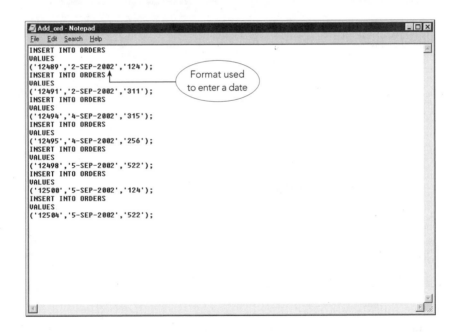

```
Add_ord - Notepad
File  Edit  Search  Help
INSERT INTO ORDERS
VALUES
('12489','2-SEP-2002','124');
INSERT INTO ORDERS
VALUES
('12491','2-SEP-2002','311');
INSERT INTO ORDERS
VALUES
('12494','4-SEP-2002','315');
INSERT INTO ORDERS
VALUES
('12495','4-SEP-2002','256');
INSERT INTO ORDERS
VALUES
('12498','5-SEP-2002','522');
INSERT INTO ORDERS
VALUES
('12500','5-SEP-2002','124');
INSERT INTO ORDERS
VALUES
('12504','5-SEP-2002','522');
```

Format used
to enter a date

Figure 2.19 contains a CREATE TABLE command for the PART table, and Figure 2.20 contains the appropriate INSERT commands to load the PART table with data.

FIGURE 2.19   CREATE TABLE command for PART table

```
cre_part.SQL - Notepad
File  Edit  Search  Help

CREATE TABLE PART
(PART_NUMBER CHAR(4) PRIMARY KEY,
PART_DESCRIPTION CHAR(12),
UNITS_ON_HAND DECIMAL(4,0),
ITEM_CLASS CHAR(2),
WAREHOUSE_NUMBER CHAR(1),
UNIT_PRICE DECIMAL(6,2) )
;
```

PART_NUMBER
is the primary key

FIGURE 2.20   INSERT commands for PART table

```
Add_part - Notepad
File  Edit  Search  Help

INSERT INTO PART
VALUES
('AX12','Iron',104,'HW','3',24.95);
INSERT INTO PART
VALUES
('AZ52','Dartboard',20,'SG','2',12.95);
INSERT INTO PART
VALUES
('BA74','Basketball',40,'SG','1',29.95);
INSERT INTO PART
VALUES
('BH22','Cornpopper',95,'HW','3',24.95);
INSERT INTO PART
VALUES
('BT04','Gas Grill',11,'AP','2',149.99);
INSERT INTO PART
VALUES
('BZ66','Washer',52,'AP','3',399.99);
INSERT INTO PART
VALUES
('CA14','Griddle',78,'HW','3',39.99);
INSERT INTO PART
VALUES
('CB03','Bike',44,'SG','1',299.99);
INSERT INTO PART
VALUES
('CX11','Blender',112,'HW','3',22.95);
INSERT INTO PART
VALUES
('CZ81','Treadmill',68,'SG','2',349.95);
```

Figure 2.21 contains a CREATE TABLE command for the ORDER_LINE table. Notice the way the primary key is defined when the primary key consists of more than one column. Figure 2.22 contains the appropriate INSERT commands to load the ORDER_LINE table with data.

FIGURE 2.21 CREATE TABLE command for ORDER_LINE

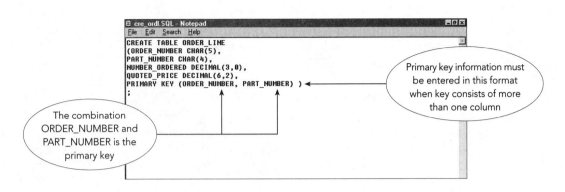

```
cre_ordl.SQL - Notepad
File  Edit  Search  Help
CREATE TABLE ORDER_LINE
(ORDER_NUMBER CHAR(5),
PART_NUMBER CHAR(4),
NUMBER_ORDERED DECIMAL(3,0),
QUOTED_PRICE DECIMAL(6,2),
PRIMARY KEY (ORDER_NUMBER, PART_NUMBER) )
;
```

Primary key information must be entered in this format when key consists of more than one column

The combination ORDER_NUMBER and PART_NUMBER is the primary key

FIGURE 2.22 INSERT commands for ORDER_LINE table

```
Add_ordl - Notepad
File  Edit  Search  Help
INSERT INTO ORDER_LINE
VALUES
('12489','AX12',11,21.95);
INSERT INTO ORDER_LINE
VALUES
('12491','BT04',1,149.99);
INSERT INTO ORDER_LINE
VALUES
('12491','BZ66',1,399.99);
INSERT INTO ORDER_LINE
VALUES
('12494','CB03',4,279.99);
INSERT INTO ORDER_LINE
VALUES
('12495','CX11',2,22.95);
INSERT INTO ORDER_LINE
VALUES
('12498','AZ52',2,12.95);
INSERT INTO ORDER_LINE
VALUES
('12498','BA74',4,24.95);
INSERT INTO ORDER_LINE
VALUES
('12500','BT04',1,149.99);
INSERT INTO ORDER_LINE
VALUES
('12504','CZ81',2,325.99);
```

■ Describing a Table

The CREATE TABLE command clearly shows the structure of the table. The command indicates all the columns, data types, and lengths of the columns. The CREATE TABLE command also indicates which columns cannot accept nulls.

When you work with a table, you might not have access to the CREATE TABLE command that was originally used to create the table. For example, someone other than you might have created the table, or perhaps you created the table several months ago but did not save the command. You still might want to examine the table's structure, however, to see details concerning the columns in the table. Each DBMS gives you a way of examining these details. In Oracle, you can use the DESCRIBE command, as shown in Figure 2.23, which lists all the columns in the SALES_REP table and their corresponding data types.

FIGURE 2.23 DESCRIBE command for SALE_REP table

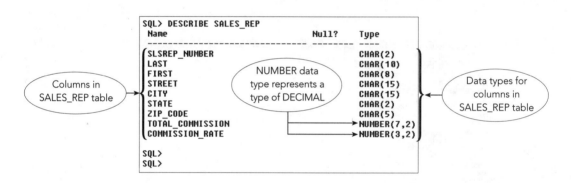

Note: In Figure 2.23, Oracle uses the word NUMBER instead of DECIMAL. Also, no semicolon is necessary at the end of the command because this is not a standard SQL command.

In this chapter, you learned how to create and run SQL commands, create tables, and add rows to these tables. In Chapter 3, you will use SQL commands to create and run queries that retrieve data from the tables and create queries that involve data from a single table. In Chapter 4, you will learn how to create queries that involve multiple tables.

SUMMARY

1. A relational database is a collection of related tables. An entity is a person, place, thing, or event. Tables in the database are entities. An attribute is a property of an entity. Attributes are the columns in the tables. A relationship is the association between tables in the database. Relationships are defined by how the data is related in columns in the tables.

2. Use the CREATE TABLE command to create a table by typing the table name and then listing, in parentheses, the columns in the table.

3. The primary key of a table is the column, or collection of columns, that uniquely identifies a given row in the table. You usually identify primary keys by underlining the column name(s).

4. You can use the editing commands listed in Table 2.1 to edit the current SQL command in the command buffer.

5. Use the DROP TABLE command to delete a table and all its data from the database.

6. The possible number data types are INTEGER, SMALLINT, DECIMAL, CHAR, and DATE.

7. A null data value (or null) is a special value that is used when the actual value for a column is unknown or unavailable. The NOT NULL command is used to identify columns that cannot accept null values, such as the table's primary key.

8. Use the INSERT command to load data into a table.

9. Use the SELECT command to view the data in a table.

10. Use the UPDATE command to change the value in a column.

11. Use the DELETE command to delete a row in a table.

12. You can use an editor, such as Notepad, to create commands that create tables and add rows to them. To run a saved command, type @ followed by the name of the file.

13. You can use the DESCRIBE command to describe the layout of a table.

■ EXERCISES (Premiere Products)

Note: If you are using Oracle for these exercises and wish to print a copy of your commands and results, you need to save the commands in a file. To do this, type SPOOL followed by the name of the file you are creating, and then press the Enter key. All the commands you enter from that point will be saved in the file that you named. For example, to save the commands and results to a file named CHAPTER2.SQL on drive A, type the following command before beginning your work:

```
SPOOL A:CHAPTER2.SQL
```

When you have finished entering and running your commands, type SPOOL OFF, and then press the Enter key to stop saving commands to the file. Then start any program that can open text files, open the file that you saved, and print it using the Print command on the File menu.

1. Use the CREATE TABLE command to create all the tables in the Premiere Products database. The CREATE TABLE commands you need are shown in Figures 2.3, 2.15, 2.17, 2.19, and 2.21.

2. Add all the sales reps shown in Figure 2.1 to the SALES_REP table using the INSERT command.

3. Add the customers shown in Figure 2.1 to the CUSTOMER table.

4. Add the orders shown in Figure 2.1 to the ORDERS table.

5. Add the order line data shown in Figure 2.1 to the ORDER_LINE table.

6. Add the part information shown in Figure 2.1 to the PART table.

■ EXERCISES (Henry Books)

1. Use the CREATE TABLE command to create all the tables in the Henry Books database. The information you need appears in Figure 2.24.

2. Add the branch information shown in Figure 1.4 (in Chapter 1) to the BRANCH table using the INSERT command.

3. Add the publisher information shown in Figure 1.4 to the PUBLISHER table.

4. Add the author information shown in Figure 1.4 to the AUTHOR table.

5. Add the book information shown in Figure 1.5 to the BOOK table.

6. Add the author and book information shown in Figure 1.6 to the WROTE table.

7. Add the inventory information shown in Figure 1.6 to the INVENT table.

8. Why is the BOOK_CODE column a character column instead of a decimal column?

FIGURE 2.24   Table layouts for Henry Books database

BRANCH

Column	Type	Length	Decimal Places	Nulls Allowed?	Description
BRANCH_NUMBER	Char	1		No	Branch number (key)
BRANCH_NAME	Char	20			Branch name
BRANCH_LOCATION	Char	20			Branch location
NUMBER_EMPLOYEES	Decimal	2	0		Number of employees

PUBLISHER

Column	Type	Length	Decimal Places	Nulls Allowed?	Description
PUBLISHER_CODE	Char	2		No	Publisher code (key)
PUBLISHER_NAME	Char	20			Publisher name
PUBLISHER_CITY	Char	20			Publisher city
PUBLISHER_STATE	Char	2			Publisher state

AUTHOR

Column	Type	Length	Decimal Places	Nulls Allowed?	Description
AUTHOR_NUMBER	Char	2		No	Author number (key)
AUTHOR_LAST	Char	20			Author last name
AUTHOR_FIRST	Char	20			Author first name

BOOK

Column	Type	Length	Decimal Places	Nulls Allowed?	Description
BOOK_CODE	Char	4		No	Book code (key)
BOOK_TITLE	Char	30			Book title
PUBLISHER_CODE	Char	2			Publisher code
BOOK_TYPE	Char	3			Book type
BOOK_PRICE	Decimal	4	2		Book price
PAPERBACK	Char	1			Paperback (Y, N)

FIGURE 2.24   Table layouts for Henry Books database (continued)

WROTE

Column	Type	Length	Decimal Places	Nulls Allowed?	Description
BOOK_CODE	Char	4		No	Book code (key)
AUTHOR_NUMBER	Char	2		No	Author number (key)
SEQUENCE_NUMBER	Decimal	1	0		Sequence number

INVENT

Column	Type	Length	Decimal Places	Nulls Allowed?	Description
BOOK_CODE	Char	4		No	Book code (key)
BRANCH_NUMBER	Char	1		No	Branch number (key)
UNITS_ON_HAND	Decimal	1	0		Units on hand

CHAPTER 3

Single-Table Queries

OBJECTIVES

- Retrieve data from a database using SQL commands

- Use compound conditions

- Use computed columns

- Use the SQL LIKE operator

- Use the SQL IN operator

- Sort data using the ORDER BY command

- Sort data using multiple keys and in ascending and descending order

- Use SQL functions

- Use nested subqueries

- Group data using the GROUP BY command

- Select individual groups using the HAVING clause

- Retrieve columns with null values

Introduction

In this chapter, you will learn about the SQL SELECT command that is used to retrieve data in a database. You will examine ways to sort data and use SQL functions to count rows and calculate totals. You also will learn about a special feature of SQL that lets you nest SELECT commands; that is, one SELECT command placed inside another. Finally, you will learn how to group rows that have matching values in some column.

■ Simple Queries

One of the most important features of a database management system is its ability to answer a wide variety of questions concerning the data in the database. When you need to find data that answers a specific question, the question is called a query. A **query** is simply a question represented in a way that the DBMS can understand.

In SQL, you use the SELECT command to query a database. The basic form of a SQL SELECT command is simple: it appears as SELECT-FROM. First, you type the word SELECT and then you list the columns that you wish to include in the query. This portion of the command is called the **SELECT clause**. Next, you type the word FROM followed by the name of the table that contains the data you need to query. This portion of the command is called the **FROM clause**. Finally, when needed, you include the word WHERE followed by any conditions (restrictions) that apply to the data you want to retrieve. This optional portion of the command is called the **WHERE clause**. For example, if you needed to retrieve the records for only those customers having a $1,000 credit limit, the WHERE clause would include a condition specifying that the value in the CREDIT_LIMIT column must be 1000 (CREDIT_LIMIT = 1000).

Remember, there are no special formatting rules for constructing SQL commands. In this book, the FROM command and the WHERE command (when it is used) appear on separate lines only to make the commands more readable and understandable.

Note: The various implementations of SQL differ on exactly how they display the query output. For example, the column headings and number formats may differ. The SQL implementation used in this book truncates (shortens) the column headings to fit the width of the column data. Although your output should contain the same data that appears in the figures in this book, its format could differ slightly.

Retrieve Certain Columns and All Rows

You can write a command to retrieve specified columns and all rows from the table, as shown in Example 1.

EXAMPLE 1 : List the customer number, last name, first name, and balance of every customer.

Because you need to list *all* customers, you do not need to include a WHERE clause (in other words, there are no restrictions). The query appears in Figure 3.1.

FIGURE 3.1 SELECT command to select certain columns

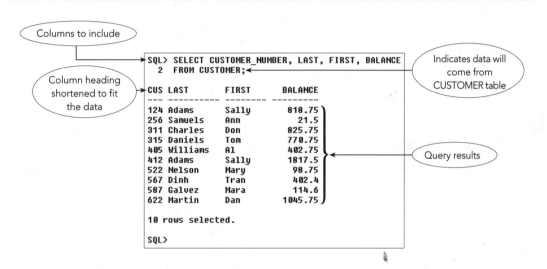

Retrieve All Columns and All Rows

You can use the same type of command as shown in Example 1 to retrieve all columns and all rows from the table. However, as Example 2 illustrates, you can use a shortcut to do this.

EXAMPLE 2 : List the complete PART table.

Instead of listing every column name after the SELECT command, you can use an asterisk (*) to indicate that you want to list all columns. The result will list all columns in the order in which you described them to the system when you created the table. If you wanted the columns listed in a different order, you would type the column names in the order in which you wanted them to appear in the query output. In this case, assuming that the default order is appropriate, you can use the query shown in Figure 3.2.

FIGURE 3.2
· · · · · · · · · · · · · · · · ·
SELECT
command to
select all
columns

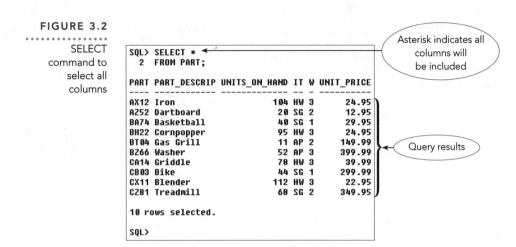

```
SQL>  SELECT *
  2   FROM PART;

PART PART_DESCRIP UNITS_ON_HAND IT W UNIT_PRICE
---- ------------ ------------- -- - ----------
AX12 Iron                   104 HW 3      24.95
A252 Dartboard               20 SG 2      12.95
BA74 Basketball              40 SG 1      29.95
BH22 Cornpopper              95 HW 3      24.95
BT04 Gas Grill               11 AP 2     149.99
B266 Washer                  52 AP 3     399.99
CA14 Griddle                 78 HW 3      39.99
CB03 Bike                    44 SG 1     299.99
CX11 Blender                112 HW 3      22.95
C281 Treadmill               68 SG 2     349.95

10 rows selected.

SQL>
```

Asterisk indicates all columns will be included

Query results

Use of the WHERE Clause—Simple Conditions

You can use the WHERE clause to retrieve records that satisfy some condition, as shown in Example 3.

EXAMPLE 3   What is the name of customer number 124?

You can use the WHERE clause to restrict the query output to customer number 124, as shown in Figure 3.3. Notice that you must enclose data for character columns, such as the CUSTOMER_NUMBER column, in single quotation marks.

FIGURE 3.3
· · · · · · · · · · · · · · · · ·
SELECT
command with
a condition

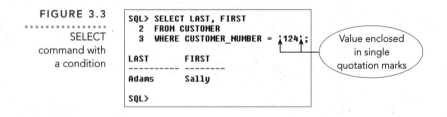

```
SQL>  SELECT LAST, FIRST
  2   FROM CUSTOMER
  3   WHERE CUSTOMER_NUMBER = '124';

LAST        FIRST
----------  --------
Adams       Sally

SQL>
```

Value enclosed in single quotation marks

The condition in the preceding WHERE clause is called a simple condition. A **simple condition** has the form: column name, comparison operator, and then either another column name or a value. The available comparison operators appear in Table 3.1. Note that there are two different operators for not equal to (< > and !=). You must use the one that is right for your implementation of SQL. (If you use the wrong operator, your system will not accept it.)

TABLE 3.1

· · · · · · · · · · · · · · · · · · ·

Comparison
operators

Comparison Operator	Description
=	Equal to
<	Less than
>	Greater than
<=	Less than or equal to
>=	Greater than or equal to
< >	Not equal to (used by most implementations of SQL)
!=	Not equal to (used by some implementations of SQL)

EXAMPLE 4 ⋮ Find the customer number for every customer whose last name is Adams.

The only difference between this example and the previous one is that in Example 3, there could not be more than one row in the answer. (Because the CUSTOMER_NUMBER column is the primary key of the CUSTOMER table, there can be only one customer whose number matches the number in the WHERE clause.) In Example 4, the results can, and do, contain more than one row, as shown in Figure 3.4.

FIGURE 3.4

· · · · · · · · · · · · · · · · ·

SELECT
command with a
condition that
retrieves
multiple rows

```
SQL> SELECT CUSTOMER_NUMBER
  2  FROM CUSTOMER
  3  WHERE LAST = 'Adams';

CUS
---
124
412

SQL>
```

Note: In general, SQL is not case sensitive. You can use uppercase or lowercase letters in any way you choose. The one important exception, however, occurs when you include values within quotation marks, as you do when entering conditions. Make sure to use the correct case for these values. For example, if you type WHERE LAST 'adams', SQL will not select any rows if the stored value is "Adams".

EXAMPLE 5 ⋮ Find the customer number, last name, first name, and current balance for every customer whose balance exceeds the credit limit.

A simple condition also can involve a comparison of two columns, as shown in Figure 3.5. The WHERE clause uses a comparison operator to find those rows in which the credit limit is greater than the balance.

FIGURE 3.5

· · · · · · · · · · · · · · · · ·

SELECT
command
involving a
comparison

```
SQL> SELECT CUSTOMER_NUMBER, LAST, FIRST, BALANCE
  2  FROM CUSTOMER
  3  WHERE BALANCE > CREDIT_LIMIT;

CUS LAST        FIRST      BALANCE
--- ---------- --------- ---------
315 Daniels    Tom         770.75
622 Martin     Dan        1045.75

SQL>
```

Compound Conditions

The conditions you have seen so far are called simple conditions. The next examples require compound conditions. You can form **compound conditions** by connecting two or more simple conditions using the AND, OR, and NOT operators. When the AND operator connects simple conditions, all the simple conditions must be true in order for the compound condition to be true. When the OR operator connects simple conditions, the compound condition will be true whenever any one of the simple conditions is true. Preceding a condition by the NOT operator reverses the truth of the original condition. For example, if the original condition is true, the new condition will be false; if the original condition is false, the new one will be true.

EXAMPLE 6 ⋮ List the description of every part that is in warehouse number 3 and that has more than 100 units on hand.

In this example, you need to retrieve those parts that meet *both* conditions—the warehouse number is equal to 3 *and* the number of units on hand is greater than 100. To find the answer, you must form a compound condition using the AND operator, as shown in Figure 3.6. The query examines the data in the PART table in the Premiere Products database and lists the parts that are located in warehouse number 3 and for which there are more than 100 units on hand.

FIGURE 3.6

· · · · · · · · · · · · · · · · ·

SELECT
command
involving an
AND condition

```
SQL> SELECT PART_DESCRIPTION
  2  FROM PART
  3  WHERE WAREHOUSE_NUMBER = '3'
  4  AND UNITS_ON_HAND > 100;

PART_DESCRIP
------------
Iron
Blender

SQL>
```

For readability, each of the simple conditions appears on a separate line. Some people prefer to put the conditions on the same line with parentheses around each simple condition, as shown in Figure 3.7. These two methods accomplish the same thing. In this text, simple conditions will appear on separate lines and without parentheses.

```
SQL> SELECT PART_DESCRIPTION
  2  FROM PART
  3  WHERE (WAREHOUSE_NUMBER = '3') AND (UNITS_ON_HAND > 100);

PART_DESCRIP
------------
Iron
Blender

SQL>
```

EXAMPLE 7 : List the description of every part that is in warehouse number 3 or that has more than 100 units on hand.

In this example, you want those parts for which the warehouse number is equal to 3 *or* the number of units on hand is greater than 100, *or* both. To do this you form a compound condition using the OR operator, as shown in Figure 3.8.

```
SQL> SELECT PART_DESCRIPTION
  2  FROM PART
  3  WHERE WAREHOUSE_NUMBER = '3'
  4  OR UNITS_ON_HAND > 100;

PART_DESCRIP
------------
Iron
Cornpopper
Washer
Griddle
Blender

SQL>
```

EXAMPLE 8 : List the description of every part that is not in warehouse number 3.

For this example, you could use a simple condition with the conditional operator for *not equal to* (WHERE WAREHOUSE_NUMBER != '3'). As an alternative, you could use the EQUAL operator (=) in the condition, and precede the entire condition with the NOT operator, as shown in Figure 3.9.

FIGURE 3.9

SELECT com-
mand involving a
NOT condition

```
SQL> SELECT PART_DESCRIPTION
  2   FROM PART
  3   WHERE NOT (WAREHOUSE_NUMBER = '3');

PART_DESCRIP
------------
Dartboard
Basketball
Gas Grill
Bike
Treadmill

SQL>
```

You do not need to enclose the condition WAREHOUSE_NUMBER = '3' in parentheses but doing so makes the command more readable. Note that by phrasing the condition in this form, you avoid the problem of determining whether your implementation of
SQL uses the < > or != not equal to operator.

Use of BETWEEN

Example 9 requires a compound condition to determine the answer.

EXAMPLE 9 ⋮ List the customer number, last name, first name, and balance for every
customer whose balance is between $500 and $1,000.

You can use the SELECT command and the AND operator as shown in Figure 3.10
to retrieve the data.

FIGURE 3.10

SELECT
command
involving an AND
condition on a
single column

```
SQL> SELECT CUSTOMER_NUMBER, LAST, FIRST, BALANCE
  2   FROM CUSTOMER
  3   WHERE BALANCE >= 500 AND BALANCE <= 1000;

CUS LAST        FIRST      BALANCE
--- ---------- --------- ---------
124 Adams       Sally       818.75
311 Charles     Don         825.75
315 Daniels     Tom         770.75

SQL>
```

An alternative to this approach uses the BETWEEN operator, as shown in Figure 3.11.

FIGURE 3.11
·················
SELECT
command
involving a
BETWEEN
condition

```
SQL> SELECT CUSTOMER_NUMBER, LAST, FIRST, BALANCE
  2  FROM CUSTOMER
  3  WHERE BALANCE BETWEEN 500 AND 1000;

CUS LAST        FIRST      BALANCE
--- ---------- -------- ---------
124 Adams       Sally        818.75
311 Charles     Don          825.75
315 Daniels     Tom          770.75

SQL>
```

The BETWEEN operator is not an essential feature of SQL; you have just seen that you can arrive at the same answer without it. Using the BETWEEN operator, however, does make certain SELECT commands simpler.

Use of Computed Columns

You can use computed columns in SQL queries. A **computed column** is a column that does not exist in the database but can be computed using data in the existing columns. Computations can involve any arithmetic operator shown in Table 3.2.

TABLE 3.2
·················
Arithmetic
operators

Arithmetic Operator	Description
+	Addition
-	Subtraction
*	Multiplication
/	Division

EXAMPLE 10   Find the customer number, last name, first name, and available credit for every customer who has a credit limit of at least $1,500.

There is no column for available credit in the Premiere Products database. However, you can compute the available credit from two columns that are present: CREDIT_LIMIT and BALANCE. To compute the available credit, you use the expression AVAILABLE_CREDIT = CREDIT_LIMIT - BALANCE, as shown in Figure 3.12.

FIGURE 3.12
· · · · · · · · · · · · · · · · · ·
SELECT
command
involving a
computed
column

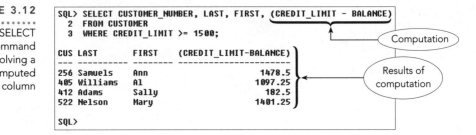

```
SQL> SELECT CUSTOMER_NUMBER, LAST, FIRST, (CREDIT_LIMIT - BALANCE)
  2  FROM CUSTOMER
  3  WHERE CREDIT_LIMIT >= 1500;

CUS LAST          FIRST    (CREDIT_LIMIT-BALANCE)
--- ---------     -------- ----------------------
256 Samuels       Ann                      1478.5
405 Williams      Al                      1097.25
412 Adams         Sally                     182.5
522 Nelson        Mary                    1401.25

SQL>
```

Computation

Results of
computation

The parentheses around the calculation (CREDIT_LIMIT - BALANCE) are not essential but improve readability.

Note: Some implementations of SQL use special headings for computed columns, such as SUM1 or COUNT2, or allow you to assign computed columns a name of your choice.

EXAMPLE 11 : Find the customer number, last name, first name, and available credit for every customer who has at least $1,000 of available credit.

You can use computed columns in comparisons, as shown in Figure 3.13. Again the parentheses around the calculation (CREDIT_LIMIT - BALANCE) are used to improve readability.

FIGURE 3.13
· · · · · · · · · · · · · · · · ·
SELECT
command with a
computation in
the condition

```
SQL> SELECT CUSTOMER_NUMBER, LAST, FIRST, (CREDIT_LIMIT - BALANCE)
  2  FROM CUSTOMER
  3  WHERE (CREDIT_LIMIT - BALANCE) >= 1000;

CUS LAST          FIRST    (CREDIT_LIMIT-BALANCE)
--- ---------     -------- ----------------------
256 Samuels       Ann                      1478.5
405 Williams      Al                      1097.25
522 Nelson        Mary                    1401.25

SQL>
```

Use of LIKE

In most cases, your conditions will involve exact matches, such as retrieving records for every customer whose last name is Adams. In some cases, however, exact matches will not work. For example, you might know that the desired value contains only a certain collection of characters. In such cases, you use the LIKE operator with a wildcard symbol, as shown in Example 12.

EXAMPLE 12 : List the customer number, last name, first name, and complete address of every customer who lives on Pine; that is, whose address contains the letters "Pine".

All you know is that the addresses that you want contain a certain collection of characters ("Pine") somewhere within them, but you don't know where. Fortunately, SQL has a facility that you can use in this situation. You can use the percentage sign (%) as a wildcard to represent any collection of characters. The condition LIKE '%Pine%', as shown in Figure 3.14, will retrieve information for every customer whose street contains some collection of characters, followed by the letters "Pine", followed by some other characters. Note that this query also would retrieve information for a customer whose address was "123 Pinell," because "Pinell" also contains the letters "Pine".

FIGURE 3.14   SELECT command with wildcards

Note: Another wildcard symbol is the underscore (_), which represents any individual character. For example, "T_m" represents the letter "T" followed by any single character, followed by the letter "m," and would retrieve records that include the words Tim, Tom, or T3m, for example.

Note: In a large database, you should use wildcards only when absolutely necessary. Searches involving wildcards can be extremely slow.

Use of IN

An IN clause provides a concise way of phrasing certain conditions, as Example 13 illustrates. You will see another use for the IN clause in more complex examples later in this chapter.

EXAMPLE 13 : List the customer number, last name, and first name for every customer with a credit limit of $1,000, $1,500, or $2,000.

In this query you will use the SQL IN operator to determine whether a credit limit is $1,000, $1,500, or $2,000. You also could obtain the same answer by using the condition WHERE CREDIT_LIMIT = 1000 OR CREDIT_LIMIT = 1500 OR CREDIT_LIMIT = 2000. The approach shown in Figure 3.15 is simpler. Here the IN clause contains a collection of values: 1000, 1500, and 2000. The condition is true for those rows in which the value in the CREDIT_LIMIT column is in this collection.

FIGURE 3.15
....
SELECT
command
involving an IN
condition

```
SQL> SELECT CUSTOMER_NUMBER, LAST, FIRST, CREDIT_LIMIT
  2  FROM CUSTOMER
  3  WHERE CREDIT_LIMIT IN (1000, 1500, 2000);

CUS LAST         FIRST     CREDIT_LIMIT
--- ----------   --------  ------------
124 Adams        Sally             1000
256 Samuels      Ann               1500
311 Charles      Don               1000
405 Williams     Al                1500
412 Adams        Sally             2000
522 Nelson       Mary              1500
587 Galvez       Mara              1000
622 Martin       Dan               1000

8 rows selected.

SQL>
```

■ Sorting

Recall that the order of rows in a table is immaterial to the DBMS. From a practical standpoint, when you are querying a relational database, there is no defined order in which the results are displayed. Rows can be displayed in the order in which the data was originally entered, but even this is not certain. If the order in which the data is displayed is important, you can *specifically* request that the results appear in a desired order. In SQL, you specify the results order by using the ORDER BY command.

Use of ORDER BY

You use the ORDER BY command to list data in a specific order, as shown in Example 14.

EXAMPLE 14 : List the customer number, last name, first name, and balance of every customer. Order the output in ascending (increasing) order by balance.

The column on which data is to be sorted is called a **sort key** or simply a **key**. In this case, because you need to order (sort) the output by balance, the sort key is the BALANCE column. To sort the output, use the ORDER BY clause, followed by the sort key. The query appears in Figure 3.16.

FIGURE 3.16   SELECT command to order rows

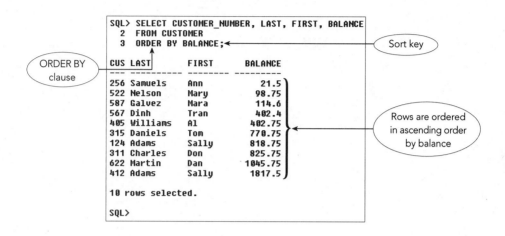

Sorting with Multiple Keys in Descending Order

Sometimes you might need to sort data by more than one key, as shown in Example 15.

EXAMPLE 15 :   List the customer number, last name, first name, and credit limit of every customer, ordered by credit limit in descending order and by last name within credit limit. The output should be sorted by credit limit in descending order. Sort the output by last name within each group of customers with the same credit limit.

This example involves two new ideas: sorting on multiple keys—CREDIT_LIMIT and LAST—and using descending order for one of the keys. If you are sorting on more than one column (such as sorting by LAST and then by CREDIT_LIMIT), the more important column (CREDIT_LIMIT) is called the **major key** (or the **primary sort key**) and the less important column (LAST) is called the **minor key** (or the **secondary sort key**). To sort on multiple keys, you list the keys in order of importance in the ORDER BY clause. To sort in descending order, you follow the name of the sort key with the DESC operator. The query appears in Figure 3.17.

FIGURE 3.17 SELECT command with multiple sort keys

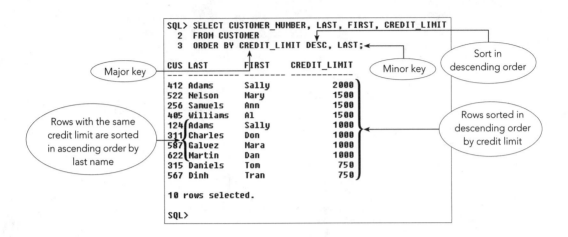

Using Functions

SQL has functions to calculate sums, averages, counts, maximum values, and minimum values. These functions and their descriptions appear in Table 3.3.

TABLE 3.3

SQL functions

Function	Description
AVG	Calculates the average value in a data series
COUNT	Determines the number of rows in a data series
MAX	Determines the maximum value in a data series
MIN	Determines the minimum value in a data series
SUM	Calculates a total of the values in a data series

Use of the COUNT Function

The COUNT function, as illustrated in Example 16, counts the number of rows in a table.

EXAMPLE 16 : How many parts are in item class HW?

For this query you need to determine the total number of rows in the PART table with the value HW in the ITEM_CLASS column. You could count the part numbers in such

56

rows or the total number of part descriptions or the number of entries in any other column. It doesn't make any difference which row you count because each count should provide the same answer. Rather than picking one column, most implementations of SQL allow you to use the asterisk (*) to represent any column, as shown in Figure 3.18.

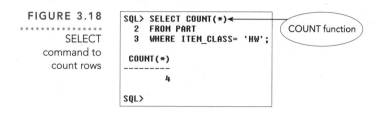

FIGURE 3.18
· · · · · · · · · · · · · · · · ·
SELECT
command to
count rows

If your implementation of SQL does not allow the use of the asterisk, you can write the query as follows:

```
SELECT COUNT(PART_NUMBER)
FROM PART
WHERE ITEM_CLASS = 'HW';
```

Use of the SUM Function

If you want to calculate the total of all customers' balances, you can use the SUM function, as illustrated in Example 17.

EXAMPLE 17 : Find the number of customers and the total of their balances.

When you use the SUM function, you must specify the column to total, and the column data type must be numeric. (How could you calculate a sum of names or addresses?) The query appears in Figure 3.19.

FIGURE 3.19
· · · · · · · · · · · · · · · · ·
SELECT
command to
calculate a count
and a sum

```
SQL> SELECT COUNT(*), SUM(BALANCE)
  2  FROM CUSTOMER;

COUNT(*) SUM(BALANCE)
-------- ------------
      10       6318.5

SQL>
```

Using the AVG, MAX, and MIN functions is similar to using SUM, except that a different statistic is calculated. Figure 3.20 shows a query that uses these three functions.

FIGURE 3.20   SELECT command with several functions

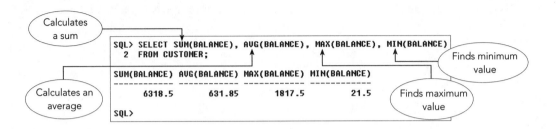

Note: When you use SUM, AVG, MAX, or MIN, SQL ignores any null value in the column; that is, null values are eliminated from the computation.

Note: Null values in numeric columns can cause strange results when statistics are computed. Suppose that the BALANCE column accepts null values, that there are currently four customers with records in the CUSTOMER table, and that their respective balances are $100, $200, $300, and null (unknown). When you calculate the average balance, most implementations will ignore the null value and obtain $200 (($100 + $200 + $300)/3). Similarly, if you calculate the total of the balances, SQL ignores the null value and calculates a total of $600. If you count the number of customers in the table, however, SQL includes the row containing the null, and the result is 4. Thus the total of the balances ($600) divided by the number of customers (4) does not equal the average balance ($200).

Use of DISTINCT

The DISTINCT operator is not a function. In some situations, however, this operator is useful when used in conjunction with the COUNT function. Before examining such a situation, you need to understand how to use the DISTINCT operator. Example 18 illustrates the most common use of DISTINCT.

EXAMPLE 18   Find the customer number of every customer who currently has an open order (that is, an order currently in the ORDERS table).

The command seems fairly simple. If a customer currently has an open order, there must be at least one row in the ORDERS table in which that customer's number appears. Thus you could use the query shown in Figure 3.21 to find the customer numbers with open orders. Note that customer numbers 124 and 522 each appear more than once in the output. The reason for this is that each customer currently has more than one open order in the ORDERS table, and thus each customer number appears more than once. Suppose you want to list each customer only once, as illustrated in Example 19.

FIGURE 3.21
·················
Results with
repeated
customer
numbers

```
SQL> SELECT CUSTOMER_NUMBER
  2  FROM ORDERS;

CUS
---
124
311
315
256
522
124
522

7 rows selected.

SQL>
```

EXAMPLE 19 : Find the customer number of every customer who currently has an open order. List each customer only once.

To ensure uniqueness, you can use the DISTINCT operator, as shown in Figure 3.22.

FIGURE 3.22
·················
Results without
repeated
customer
numbers

```
SQL> SELECT DISTINCT(CUSTOMER_NUMBER)
  2  FROM ORDERS;

CUS
---
124
256
311
315
522

SQL>
```

Now, let's examine the relationship between COUNT and DISTINCT.

EXAMPLE 20 : Count the number of customers who currently have open orders.

The query shown in Figure 3.23 shows the number of customers with records in the CUSTOMER_NUMBER column.

FIGURE 3.23
....
Count that
includes
repeated
customer
numbers

```
SQL> SELECT COUNT(CUSTOMER_NUMBER)
  2  FROM ORDERS;

COUNT(CUSTOMER_NUMBER)
----------------------
                     7

SQL>
```

QUESTION   What's wrong with the query results shown in Figure 3.23?

ANSWER   The answer, 7, is the result of counting the customers who have open orders multiple times—once for each separate order they currently have on file. The result counts each customer number and does not eliminate redundant customer numbers to provide an accurate count of the number of customers.

To solve this problem, you can use the DISTINCT operator, as shown in Figure 3.24.

FIGURE 3.24
....
Count without
repeated
customer
numbers

```
SQL> SELECT COUNT(DISTINCT(CUSTOMER_NUMBER))
  2  FROM ORDERS;

COUNT(DISTINCT(CUSTOMER_NUMBER))
--------------------------------
                               5

SQL>
```

■ Nesting Queries

Sometimes obtaining the results you need is a two-step process (or more), as shown in the next two examples.

EXAMPLE 21 : What is the largest credit limit given to any customer of sales rep 06?

Use the MAX function, as shown in Figure 3.25.

EXAMPLE 22 : Display the customer number, last name, and first name of every customer in the Premiere Products database who has the credit limit found
in Example 21.

You need to find all customers with this credit limit, not just customers of sales rep 06.
After viewing the answer to the previous example (1500), you could use the command
shown in Figure 3.26.

FIGURE 3.26
. . . .
Query using
previous result

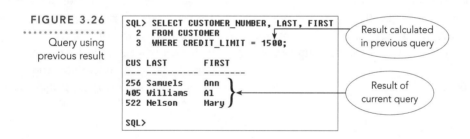

Subqueries

It is possible to place one query inside another. The inner query is called a **subquery** and
it is evaluated first. Then the outer query can use the results of the subquery to find its
results.

EXAMPLE 23 : Find the answer to Examples 21 and 22 in one step.

You can find the same result as in the previous two examples, but in a single step, by
using a subquery. The form for subqueries is shown in Figure 3.27. The portion in parentheses is the subquery. This subquery is evaluated first, producing a temporary table. The
table is used strictly in evaluating the query and is not available to the user or displayed.

This temporary table is deleted after the evaluation of the query is complete. In this example, the temporary table has only a single column MAX(CREDIT_LIMIT) and a single row containing the number 1500.

FIGURE 3.27

Using IN and a subquery

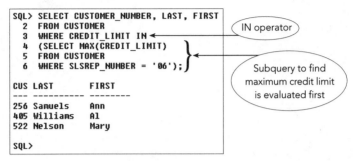

```
SQL> SELECT CUSTOMER_NUMBER, LAST, FIRST
  2  FROM CUSTOMER
  3  WHERE CREDIT_LIMIT IN
  4  (SELECT MAX(CREDIT_LIMIT)
  5  FROM CUSTOMER
  6  WHERE SLSREP_NUMBER = '06');

CUS LAST        FIRST
--- ----------  --------
256 Samuels     Ann
405 Williams    Al
522 Nelson      Mary

SQL>
```

IN operator

Subquery to find maximum credit limit is evaluated first

The outer query is evaluated next. The outer query will retrieve the customer number, last name, and first name of every customer whose credit limit is in the temporary table result produced by the subquery. Because that table contains only the maximum credit limit for the customers of sales rep 06, you obtain the desired list of customers. Incidentally, because the subquery in this case produces a table containing only a single value (the maximum credit limit), this query also could have been formulated as shown in Figure 3.28, where you are requesting those customers whose credit limit is *equal to* the one unique credit limit obtained by the subquery. In general, unless you know that the subquery *must* produce a single value, the previous command using IN is the correct one to use.

FIGURE 3.28

Query using an EQUAL condition and a subquery

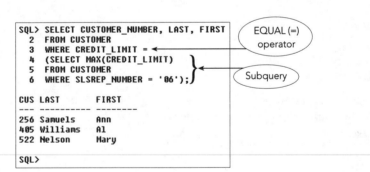

```
SQL> SELECT CUSTOMER_NUMBER, LAST, FIRST
  2  FROM CUSTOMER
  3  WHERE CREDIT_LIMIT =
  4  (SELECT MAX(CREDIT_LIMIT)
  5  FROM CUSTOMER
  6  WHERE SLSREP_NUMBER = '06');

CUS LAST        FIRST
--- ----------  --------
256 Samuels     Ann
405 Williams    Al
522 Nelson      Mary

SQL>
```

EQUAL (=) operator

Subquery

In Example 24, the subquery again produces a single value, but it uses the > operator rather than the = operator in the condition.

EXAMPLE 24 : List the customer number, first name, last name, and balance for every customer whose balance is greater than the average balance.

In this case, you use a subquery to obtain the average balance. Because this subquery produces a single number, you can compare each customer's balance with this number, as shown in Figure 3.29.

FIGURE 3.29
· · · · · · · · · · · · · · · · ·
Query using
greater than
operator and a
subquery

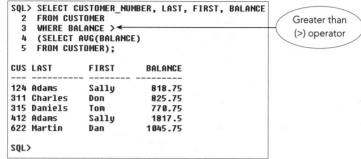

```
SQL> SELECT CUSTOMER_NUMBER, LAST, FIRST, BALANCE
  2  FROM CUSTOMER
  3  WHERE BALANCE >
  4  (SELECT AVG(BALANCE)
  5  FROM CUSTOMER);

CUS LAST        FIRST      BALANCE
--- ---------- -------- ---------
124 Adams       Sally      818.75
311 Charles     Don        825.75
315 Daniels     Tom        770.75
412 Adams       Sally      1817.5
622 Martin      Dan       1045.75

SQL>
```

Greater than
(>) operator

Note: SQL will not allow you to use the condition BALANCE > AVG(BALANCE) in the WHERE clause. You must use a subquery to obtain the average balance. Then you can use the results of the subquery in your condition as illustrated in Figure 3.29.

▌Grouping

Grouping creates groups of rows that share some common characteristic. If customers were grouped by credit limit, for example, the first group would contain customers with a $750 credit limit, the second group would contain customers with a $1,000 credit limit, and so on. If order lines were grouped by order number, those order lines for order 12489 would form one group, those for order 12491 would form a second group, and so on.

When you group records, any calculations indicated in the SELECT command are performed for the entire group. For example, if customers are grouped by credit limit and the query requests the average balance, the results include the average balance for the group of customers with a $750 credit limit, the average balance for the group with a $1,000 credit limit, and so on. The following examples illustrate this process.

Using GROUP BY

The GROUP BY command allows you to group data in a particular order, and then to calculate statistics if desired, as shown in Example 25.

To obtain the total for a particular order, first you multiply the number of units ordered by the quoted price for each order line in the order. Then you add these results. You repeat this process for each order. To calculate totals for individual orders, you cannot simply include SUM(NUMBER_ORDERED * QUOTED_PRICE) in the query. This expression will calculate the grand total for all order lines; the grand total will not be broken down by order. To calculate individual totals you can use the GROUP BY command. In this case, the command GROUP BY ORDER_NUMBER groups the order lines for each order; that is, all order lines with the same order number form a group. Any statistics, such as totals, requested in the SELECT command are calculated for each group. It is important to note that the GROUP BY command does not sort the data in a particular order. You must use the ORDER BY command to sort data. Assuming that the report is to be ordered by order number, you can use the command shown in Figure 3.30.

FIGURE 3.30 Grouping by a column

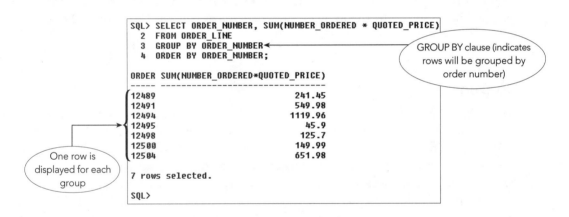

When rows are grouped, one line of output is produced for each group. The only data that can be displayed is statistics calculated for the group or columns whose values are the same for all rows in a group.

QUESTION Would it be appropriate to display the order number?

ANSWER Yes, because the output is grouped by order number. The order number in one row in a group must be the same as the order number in any other row in the group.

QUESTION   Would it be appropriate to display a part number?

ANSWER   No, because the part number varies from one row in a group to another.
            (The same order can contain many parts.) Thus, SQL would not be able to
            determine which part number to display for the group. The system dis-
            plays an error message if you attempt to list a part number.

Using HAVING

The HAVING command is used for groups, as shown in Example 26.

EXAMPLE 26 :   List the total for those orders over $200.

   This example is similar to the previous one; the only difference is the restriction to display
totals for orders that are greater than $200. This restriction does not apply to individual rows
but rather to *groups*. Because the WHERE command applies only to rows, it is not the appropriate command to accomplish the kind of selection that is required. Fortunately, the HAVING
clause does for groups what the WHERE command does for rows. In Figure 3.31, the row
created for a group will be displayed only if the sum calculated for the group is larger than $200;
in addition, all groups will be ordered by order number.

FIGURE 3.31   Query using a HAVING clause

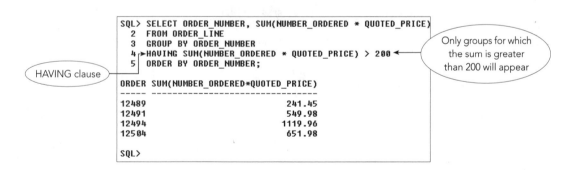

HAVING vs. WHERE

Just as you can use the WHERE clause to limit the *rows* that are included in the result of a
SQL command, you can use the HAVING clause to limit the *groups* that are included. The
following examples illustrate the difference between these two clauses.

EXAMPLE 27 : List each credit limit and the number of customers having each
credit limit.

In order to count the number of customers who have a given credit limit, you must
group the data by credit limit, as shown in Figure 3.32.

```
SQL> SELECT CREDIT_LIMIT, COUNT(*)
  2  FROM CUSTOMER
  3  GROUP BY CREDIT_LIMIT;

CREDIT_LIMIT   COUNT(*)
------------   --------
         750          2
        1000          4
        1500          3
        2000          1

SQL>
```

EXAMPLE 28 : Repeat Example 27, but list only those credit limits held by more than
one customer.

Because this condition involves a group total, a HAVING clause is used, as shown in
Figure 3.33.

```
SQL> SELECT CREDIT_LIMIT, COUNT(*)
  2  FROM CUSTOMER
  3  GROUP BY CREDIT_LIMIT
  4  HAVING COUNT(*) > 1;

CREDIT_LIMIT   COUNT(*)
------------   --------
         750          2
        1000          4
        1500          3

SQL>
```

EXAMPLE 29 : List each credit limit and the total number of customers of sales rep 03
who have this limit.

The condition involves only rows, so using the WHERE clause is appropriate, as
shown in Figure 3.34.

FIGURE 3.34
....
Restricting the
rows to be
grouped

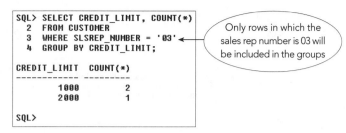

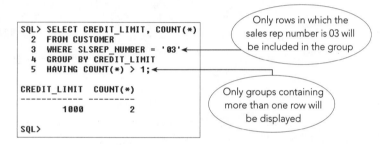

EXAMPLE 30 : Repeat Example 29, but list only those credit limits held by more than one customer.

Because the conditions involve rows and groups, you must use both a WHERE clause and a HAVING clause, as shown in Figure 3.35.

FIGURE 3.35
....
Restricting the
rows and the
groups

In Example 30, rows from the original table are considered only if the sales rep number is 03. These rows are then grouped by credit limit and the count is calculated. Only groups for which the calculated count is greater than 1 are displayed.

■ Nulls

Sometimes a condition involves a column that can be null, as illustrated in Example 31.

EXAMPLE 31 : List the customer number, last name, and first name of every customer whose street value is null (unknown).

You might expect the condition to be something like STREET = NULL. The correct format is actually STREET IS NULL, as shown in Figure 3.36. (To select a customer whose street is not null, you use the condition STREET IS NOT NULL.) In the current

Premiere Products database, no customer has a null street value; therefore no rows are retrieved in the query results.

FIGURE 3.36
················
Selecting rows
containing null
values

```
SQL> SELECT CUSTOMER_NUMBER, LAST, FIRST
  2  FROM CUSTOMER
  3  WHERE STREET IS NULL;

no rows selected

SQL>
```

In this chapter, you learned how to create queries that retrieve data from a single table by constructing an appropriate SELECT command. In the next chapter, you will learn how to create queries that involve multiple tables. The queries you created in this chapter used the clauses and operators shown in Table 3.4.

TABLE 3.4
················
SQL query
clauses and
operators

Clause or Operator	Description
AND operator	All simple conditions must be true in order for the compound condition to be true
BETWEEN operator	Specifies a range of values in a condition
DESC operator	Sorts the query results in descending order based on the column name
DISTINCT operator	Ensures uniqueness in the condition by eliminating redundant values
FROM clause	Indicates the table from which to retrieve the specified columns
GROUP BY clause	Groups rows based on the specified column
HAVING clause	Limits a condition to the groups that are included
IN operator	Finds a value in a group of values specified in the condition
IS NOT NULL operator	Finds rows that do not contain a null value in the specified column
IS NULL operator	Finds rows that contain a null value in the specified column
LIKE operator	Indicates a pattern of characters to find in a condition
NOT operator	Reverses the truth or falsity of the original condition
OR operator	The compound condition is true whenever any of the simple conditions is true
ORDER BY clause	Lists the query results in the specified order based on the column name
SELECT clause	Specifies the columns to retrieve in the query
WHERE clause	Specifies any conditions for the query

SUMMARY

1. The basic form of a SQL command is SELECT-FROM. Specify the columns to be listed after the word SELECT (or type * to select all columns), and then specify the table name that contains these columns after the word FROM. Optionally, you can include conditions after the word WHERE.

2. Simple conditions are written in the form: column name, comparison operator, column name or value. Simple conditions can involve any of the comparison operators: =, >, >=, <, <=, or < > or != (not equal to).

3. You can form compound conditions by combining simple conditions, using the operators AND, OR, or NOT.

4. Use the BETWEEN operator to indicate a range of values in a condition.

5. Use computed columns in SQL commands by using arithmetic operators and writing the computation in place of a column name.

6. To check for a value in a character column that is similar to a particular string of characters, use the LIKE clause. The

% wildcard represents any collection of characters. The _ wildcard represents any single character.

7. To check whether a column contains one of a particular set of values, use the IN clause.

8. Use the ORDER BY clause to sort data. List sort keys in order of importance. To sort in descending order, follow the sort key with DESC.

9. SQL contains the functions COUNT, SUM, AVG, MAX, and MIN.

10. To avoid duplicates, either when listing or counting values, precede the column name with the DISTINCT operator.

11. When one SQL query is placed inside another, it is called a nested query. The inner query, called a subquery, is evaluated first.

12. Use the GROUP BY clause to group data.

13. Use the HAVING clause to restrict the output to certain groups.

14. Use the phrase IS NULL in the WHERE clause to find rows containing a null value in some column.

■EXERCISES (Premiere Products)

Use SQL and Figure 1.2 in Chapter 1 to complete the following exercises.

Note: If you are using Oracle for these exercises and wish to print a copy of your commands and results, type SPOOL followed by the name of a file and then press the Enter key. All the commands from that point on are saved in the file that you named. For example, to save the commands and results to a file named CHAPTER3.SQL on drive A, type the following command before beginning your work:

```
SPOOL A:CHAPTER3.SQL
```

When you have finished, type SPOOL OFF, and then press the Enter key to stop saving commands to the file. Then start any program that opens text files, open the file that you saved, and print it using the Print command on the File menu.

1. List the part number and part description for all parts.

2. List all rows and columns for the complete SALES_REP table.

3. Find the last name and first name for every customer who has a credit limit of at least $800.

4. Give the order number for every order placed by customer number 124 on 9/05/2002.

5. List the customer number, last name, and first name for every customer represented by sales rep 03 or sales rep 12.

6. List the part number and part description for every part that is not in item class HW.

7. List the part number and part description for every part that has between 100 and 200 units on hand. Do this two ways.

8. Give the part number, part description, and on-hand value (units on hand * unit price) for each part in item class AP. (On-hand value is really units on hand * cost, but there is no cost column in the PART table.)

9. List the part number, part description, and on-hand value for each part whose on-hand value is at least $1,000.

10. List the part number and part description for every part whose item class is HW or SG. Use the IN operator in your command.

11. Find the customer number, last name, and first name for every customer whose first name begins with the letter "D".

12. List all details about all parts. Order the output by part description.

13. List all details about all parts. Order the output by part number within item class. (That is, order the output by item class and then by part number.)

14. Find out how many customers have a balance that is less than their credit limit.

15. Find the total of the balances for all customers represented by sales rep 12 who have a balance that is less than their credit limits.

16. List the part number, part description, and units on hand of all parts whose number of units on hand is more than average.

17. What is the command to determine the least expensive part in the database?

18. What is the command to determine the most expensive part in the database?

19. What is the command to determine how many customers are in the database?

20. List the sum of the balances of all customers for every sales rep. Order and group the results using the sales rep number.

21. List the item class and the sum of the value of parts on hand. Group the results by item class.

22. Find any parts with an unknown part description.

EXERCISES (Henry Books)

Use SQL and Figures 1.4 through 1.6 in Chapter 1 to complete the following exercises.

1. List the book code and book title for every book.

2. List the complete PUBLISHER table.

3. List the name of every publisher located in New York state.

4. List the name of every publisher not located in New York state.

5. List the name of every branch that has at least 10 employees.

6. List the code and title of every book whose type is HOR.

7. List the code and title of every book whose type is HOR and that is paperback.

8. List the codes and titles of all books whose type is HOR or whose publisher code is PB.

9. List the code, title, and price for each book with a price that is greater than $10 but less than $20.

10. List the book code and title of every book whose type is MYS and whose price is less than $20.

11. Customers who are part of a special program get a 15% discount on regular book prices. To determine the discounted prices, list the book code, title, and discounted price of every book. (Your calculated column should determine 85% of the current price; that is, 100% less a 15% discount.)

12. Find the name of every publisher containing the word "and".

13. List the book code and title of every book whose type is FIC, MYS, or ART. Use the IN operator in your command.

14. Repeat Exercise 13 and list the books in alphabetical order by title.

15. Repeat Exercise 13 and list the books in descending order by book code.

16. List the last and first name of every author. Order the output by last name.

17. Find out how many book types are available. (Do not include duplicates.)

18. Find out how many books are of type MYS.

19. Calculate the average price for each type of book.

20. Repeat Exercise 19, but consider only paperback books.

21. What is the command to determine the name of the most expensive book?

22. What is the command to determine how many employees Henry has?

23. What is the command to determine the branch name that employs the least people?

24. What is the command to determine the title of the most expensive book published by Pocket Books?

CHAPTER 4

Multiple-Table Queries

OBJECTIVES

- Retrieve data from more than one table by joining tables

- Use the IN and EXISTS operators to query multiple tables

- Use a subquery within a subquery

- Use an alias

- Join a table to itself

- Perform set operations (union, intersection, and difference)

- Use the ALL and ANY operators in a query

Introduction

In this chapter, you will learn how to retrieve data from two or more tables using one SQL statement. You will see how tables can be joined together and how similar results are obtained using the IN and EXISTS operators. Then you will use aliases to simplify queries and join a table to itself. You will also see how to implement the set operations of union, intersection, and difference using SQL commands. Finally, you will examine two other SQL operators: ALL and ANY.

■ Querying Multiple Tables

In Chapter 3 you learned how to retrieve data from a single table. Sometimes, however, you might need to retrieve data from two or more tables. To retrieve data from two or more tables, first you must join the tables. Then you formulate a query using the same commands that you use for single tables.

Joining Two Tables

One common way to retrieve data from more than one table is to **join** the tables together by finding rows in the two tables that have identical values in matching columns. You can join tables by using the appropriate conditions in the WHERE clause, as you will see in Example 1.

EXAMPLE 1   List the customer number, last name, and first name for every customer, together with the sales rep number, last name, and first name for the sales rep who represents each customer.

Because the customers' numbers and names are in the CUSTOMER table and the sales reps' numbers and names are in the SALES_REP table, you need to join the tables in the same SQL command so you can retrieve data from both tables. To join tables you must:

1. Indicate in the SELECT clause all columns to display.

2. List in the FROM clause all tables involved in the query.

3. Give condition(s) in the WHERE clause to restrict the data to be retrieved to only those rows that have common values in matching columns.

There can be a problem, however, when matching columns. The matching columns in this example are both named SLSREP_NUMBER: there is a column in the SALES_REP table named SLSREP_NUMBER, and a column in the CUSTOMER table that also is named SLSREP_NUMBER. If you reference the SLSREP_NUMBER column, it is not clear which table the column is from. In this case it is necessary to **qualify** the SLSREP_NUMBER column to indicate which of the columns you are referencing. You qualify columns by separating the table name and the column name with a period. For example, you would write the SLSREP_NUMBER column in the SALES_REP table as SALES_REP.SLSREP_NUMBER, and you would write the SLSREP_NUMBER column in the CUSTOMER table as CUSTOMER.SLSREP_NUMBER. The query format and results appear in Figure 4.1. Notice that because both the SALES_REP and CUSTOMER tables contain columns named LAST and FIRST, you must qualify these column names as well.

FIGURE 4.1   Joining tables

```
SQL> SELECT CUSTOMER_NUMBER, CUSTOMER.LAST, CUSTOMER.FIRST,
   2   SALES_REP.SLSREP_NUMBER, SALES_REP.LAST, SALES_REP.FIRST
   3   FROM CUSTOMER, SALES_REP
   4   WHERE CUSTOMER.SLSREP_NUMBER = SALES_REP.SLSREP_NUMBER;

CUS LAST          FIRST     SL LAST       FIRST
--- ----------    --------  -- -------    --------
124 Adams         Sally     03 Jones      Mary
412 Adams         Sally     03 Jones      Mary
622 Martin        Dan       03 Jones      Mary
256 Samuels       Ann       06 Smith      William
315 Daniels       Tom       06 Smith      William
567 Dinh          Tran      06 Smith      William
587 Galvez        Mara      06 Smith      William
311 Charles       Don       12 Diaz       Miguel
405 Williams      Al        12 Diaz       Miguel
522 Nelson        Mary      12 Diaz       Miguel

10 rows selected.

SQL>
```

Tables to include in query results

CUSTOMER table

Condition to relate the tables

SLSREP_NUMBER column

Data from CUSTOMER table

Data from SALES_REP table

. . . .

QUESTION   In the first row of output in Figure 4.1, the customer number is 124, the last name is Adams, and the first name is Sally. These values represent the first row of the CUSTOMER table. Why is the sales rep number 03, the last name of the sales rep Jones, and the first name Mary?

ANSWER   In the CUSTOMER table, the sales rep number for customer number 124 is 03. (This indicates that customer number 124 is *related* to sales rep number 03.) In the SALES_REP table, the last name of sales rep number is 03 is Jones and the first name is Mary.

. . . .

Whenever there is potential ambiguity among column names, you *must* qualify the columns involved in the query. It is permissible to qualify other columns as well, even if there is no possibility of confusion. Some people prefer to qualify all column names; in this text, however, column names will be qualified only when it is necessary to avoid confusion.

List the customer number, last name, and first name of every customer whose credit limit is $1,000 together with the sales rep number, last name, and first name of the sales rep who represents each customer.

In Example 1, you used the condition in the WHERE clause only to relate a customer with a sales rep and to join the tables. Relating a customer to a sales rep is essential in this example as well, but you also need to restrict the output to rows for those customers having a credit limit of $1,000. You can restrict the rows by using a compound condition, as shown in Figure 4.2.

FIGURE 4.2 Restricting the rows in a join

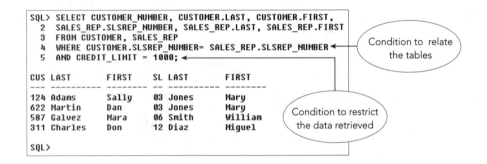

EXAMPLE 3 For every part on order, list the order number, part number, part description, number of units ordered, quoted price, and the unit price.

A part is considered to be on order if there is a row in the ORDER_LINE table in which the part appears. You can find the order number, number of units ordered, and the quoted price in the ORDER_LINE table. To find the part description and the unit price, however, you need to look in the PART table. Then you need to find rows in the ORDER_LINE table and rows in the PART table that match (rows containing the same part number). The query format and results appear in Figure 4.3.

FIGURE 4.3

Joining the
ORDER_LINE
and PART tables

```
SQL> SELECT ORDER_NUMBER, ORDER_LINE.PART_NUMBER, PART_DESCRIPTION,
  2    NUMBER_ORDERED, QUOTED_PRICE, UNIT_PRICE
  3  FROM ORDER_LINE, PART
  4  WHERE ORDER_LINE.PART_NUMBER = PART.PART_NUMBER;

ORDER PART PART_DESCRIP NUMBER_ORDERED QUOTED_PRICE UNIT_PRICE
----- ---- ------------ -------------- ------------ ----------
12489 AX12 Iron                     11        21.95      24.95
12491 BT04 Gas Grill                 1       149.99     149.99
12491 BZ66 Washer                    1       399.99     399.99
12494 CB03 Bike                      4       279.99     299.99
12495 CX11 Blender                   2        22.95      22.95
12498 AZ52 Dartboard                 2        12.95      12.95
12498 BA74 Basketball                4        24.95      29.95
12500 BT04 Gas Grill                 1       149.99     149.99
12504 CZ81 Treadmill                 2       325.99     349.95

9 rows selected.

SQL>
```

QUESTION   Can you use PART.PART_NUMBER in place of ORDER_LINE.PART_NUMBER
            in the SELECT clause?

ANSWER   Yes, because the values for these two columns match and therefore satisfy
            the condition ORDER_LINE.PART_NUMBER = PART.PART_NUMBER.

Comparison of JOIN, IN, and EXISTS

You join tables in SQL by including a condition in the WHERE clause to ensure that matching columns contain equal values (for example, ORDER_LINE.PART_NUMBER = PART.PART_NUMBER). You can obtain similar results by using either the IN operator (used in Chapter 3) or the EXISTS operator with a subquery. The choice is a matter of personal preference because either approach obtains the same results. The following examples illustrate the use of each operator.

EXAMPLE 4   Find the descriptions for every part included in order number 12491.

Because this query also involves retrieving data from the ORDER_LINE and PART tables as illustrated in Example 3, you could approach it in a similar fashion. There are two basic differences, however, between Examples 3 and 4. First, the query in Example 4 does not require as many columns; second, it involves only order number 12491. The fact that there are fewer columns means that there will be fewer attributes listed in the SELECT clause. You can restrict the query to a single order by adding the condition ORDER_NUMBER = '12491' to the WHERE clause. The query format and results appear in Figure 4.4.

```
SQL> SELECT PART_DESCRIPTION
  2  FROM ORDER_LINE, PART
  3  WHERE ORDER_LINE.PART_NUMBER = PART.PART_NUMBER
  4  AND ORDER_NUMBER = '12491';

PART_DESCRIP
------------
Gas Grill
Washer

SQL>
```

Only order lines for order number 12491 will be included

Notice that the ORDER_LINE table is listed in the FROM clause, even though you don't need to display any columns from the ORDER_LINE table. The WHERE clause contains columns from the ORDER_LINE table, so it is necessary to include the table name in the FROM clause.

Using IN

Another way to retrieve data from multiple tables in a query is to use the IN operator with a subquery. In Example 4, you could first use a subquery to find all part numbers in the ORDER_LINE table that appear in any row in which the order number is 12491. Then you could find the part description for any part whose number is in this list. The query format and results appear in Figure 4.5.

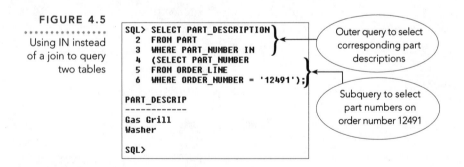

```
SQL> SELECT PART_DESCRIPTION
  2  FROM PART
  3  WHERE PART_NUMBER IN
  4  (SELECT PART_NUMBER
  5  FROM ORDER_LINE
  6  WHERE ORDER_NUMBER = '12491');

PART_DESCRIP
------------
Gas Grill
Washer

SQL>
```

Outer query to select corresponding part descriptions

Subquery to select part numbers on order number 12491

In Figure 4.5, evaluating the subquery produces a temporary table consisting of those part numbers (BT04 and BZ66) that are present in order number 12491. Executing the remaining portion of the query produces part descriptions for each part whose number is in this temporary table, in this case, Gas Grill (BT04) and Washer (BZ66).

Using EXISTS

You also can use the EXISTS operator to retrieve data from more than one table, as shown in Example 5.

EXAMPLE 5 : Find the order number and order date for every order that contains part number BT04.

This query is similar to the one in Example 4 but this time the query involves the ORDERS table and not the PART table. Here you can write the query in either of the ways just demonstrated. Using the IN operator with a subquery produces the query format and results shown in Figure 4.6. Notice that the date is displayed as two digits, which is the default in some systems, even though you entered it into the table as 2002.

FIGURE 4.6
· · · · · · · · · · · · · · · ·
Using IN to
select order
information

```
SQL> SELECT ORDER_NUMBER, ORDER_DATE
  2  FROM ORDERS
  3  WHERE ORDER_NUMBER IN
  4  (SELECT ORDER_NUMBER
  5  FROM ORDER_LINE
  6  WHERE PART_NUMBER = 'BT04');

ORDER ORDER_DAT
----- ---------
12491 02-SEP-02
12500 05-SEP-02

SQL>
```

Using the EXISTS operator provides another approach to the problem, as shown in Figure 4.7.

FIGURE 4.7
· · · · · · · · · · · · · · · ·
Using EXISTS to
select order
information

```
SQL> SELECT ORDER_NUMBER, ORDER_DATE
  2  FROM ORDERS
  3  WHERE EXISTS
  4  (SELECT *
  5  FROM ORDER_LINE
  6  WHERE ORDERS.ORDER_NUMBER = ORDER_LINE.ORDER_NUMBER
  7  AND PART_NUMBER = 'BT04');

ORDER ORDER_DAT
----- ---------
12491 02-SEP-02
12500 05-SEP-02

SQL>
```

The subquery in Figure 4.7 is the first you have seen that involves a table mentioned in the outer query; this subquery is called a **correlated subquery**. In this case, the ORDERS table, which is listed in the FROM clause of the outer query, is used in the subquery (ORDERS.ORDER_NUMBER). For this reason you need to qualify the

ORDER_NUMBER column in the subquery. You did not need to qualify the columns in the previous queries involving IN.

This query works as follows. For each row in the ORDERS table, the subquery is executed using the value of ORDERS.ORDER_NUMBER that occurs in that row. The inner query produces a list of all rows in the ORDER_LINE table in which ORDER_LINE.ORDER_NUMBER matches this value and in which the PART_NUMBER is equal to BT04. You can use the EXISTS operator in front of a subquery to create a condition that is true if one or more rows are obtained when the subquery is executed; otherwise, the condition is false.

To illustrate the process, consider order numbers 12491 and 12494 in the ORDERS table. Order number 12491 is included because a row exists in the ORDER_LINE table with this order number and part number BT04. When the subquery executes, there will be at least one row in the results, which in turn makes the EXISTS condition true. Order number 12494, however, will not be included because no row exists in the ORDER_LINE table with this order number and part number BT04. There will be no rows contained in the results of the subquery, which in turn makes the EXISTS condition false.

Using a Subquery within a Subquery

You can use **nested subqueries**, that is, a subquery within a subquery, as illustrated in Example 6.

EXAMPLE 6 : Find the order number and order date for every order that includes a part located in warehouse number 3.

One way to approach this problem is first to determine the list of part numbers in the PART table for every part located in warehouse number 3. Then you obtain a list of order numbers in the ORDER_LINE table with a corresponding part number in the part number list. Finally, you retrieve those order numbers and order dates in the ORDERS table for which the order number is in the list of order numbers obtained during the second step. The query format and results appear in Figure 4.8.

FIGURE 4.8
. . . .
Nested
subqueries
(a subquery
within a
subquery)

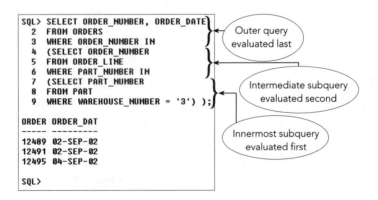

As you would expect, the queries are evaluated from the innermost query to the outermost query. The query in this example is evaluated in three steps:

1. The innermost subquery is evaluated first, producing a temporary table of part numbers for those parts located in warehouse number 3.

2. The next (intermediate) subquery is evaluated, producing a second temporary table with a list of order numbers. Each order number in this collection has a row in the ORDER_LINE table for which the part number is in the temporary table produced in Step 1.

3. The outer query is evaluated last, producing the desired list of order numbers and order dates. Only those orders whose numbers are in the temporary table produced in Step 2 are included in the result.

An alternate formulation involves joining the ORDERS, ORDER_LINE, and PART tables. The query format and results appear in Figure 4.9.

FIGURE 4.9

Joining three tables

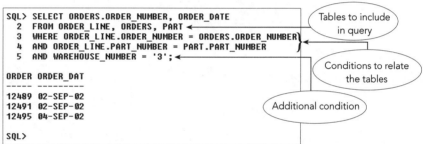

In this formulation, the conditions ORDER_LINE.ORDER_NUMBER = ORDERS.ORDER_NUMBER and ORDER_LINE.PART_NUMBER = PART.PART_NUMBER join the tables. The condition WAREHOUSE_NUMBER = '3' restricts the output to only those parts located in warehouse number 3.

The query results are correct regardless of which formulation you use. You can use whichever approach you prefer.

You might wonder whether one approach is more efficient than the other. Good mainframe systems have built-in optimizers that analyze queries to determine the best way to satisfy them. Given a good optimizer, it should not make any difference how you formulate the query. If you are using a system without such an optimizer, the formulation of a query *can* make a difference in the speed with which the query is executed. If you are working with very large databases and efficiency is a prime concern, you can consult your system's manual or try some timings yourself. Try running the same query both ways to see if you notice a difference in the speed of execution. In small databases, there should not be a significant time difference between the two approaches.

A Comprehensive Example

The query used in Example 7 involves several of the features already discussed. It illustrates all the major clauses that you can use in the SELECT command. It also illustrates the order in which these clauses must appear.

EXAMPLE 7　　　List the customer number, order number, order date, and order total for every order with a total of over $100.

The query format and results appear in Figure 4.10.

FIGURE 4.10　　Comprehensive example

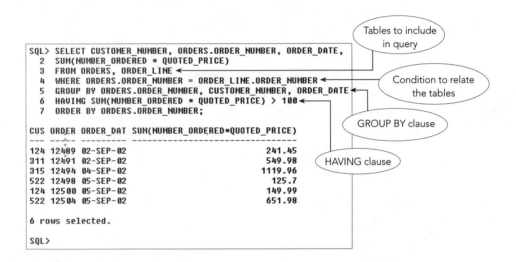

In this example, the ORDERS and ORDER_LINE tables are joined by listing both tables in the FROM clause and relating them in the WHERE clause. Selected data is sorted by ORDER_NUMBER using the ORDER BY clause. The GROUP BY clause indicates that the data is to be grouped by order number, customer number, and order date. For each group, the SELECT statement indicates that the customer number, order number, order date, and order total (SUM(NUMBER_ORDERED * QUOTED_PRICE)) will be displayed. Not all groups will be displayed, however. The HAVING clause indicates that only those groups whose SUM(NUMBER_ORDERED * QUOTED_PRICE) is greater than 100 will be displayed.

The order number, customer number, and order date are unique to each order. Thus, it would seem that merely grouping by order number would be sufficient. Most implementations of SQL, including Oracle, still require that both customer number and order date be listed in the GROUP BY clause. Recall that the SELECT statement can include statistics calculated for only the groups or columns whose values are known to be the same for each row in a group. By stating that the data is to be grouped by order number, customer number, and order date, you tell the system that the values in these columns must be the same for each row in a group. A more sophisticated implementation would realize that given the structure of this database, grouping by order number alone is sufficient to ensure the uniqueness of both customer number and order date.

■ Using an Alias

When tables are listed in the FROM clause, you can give each table an **alias,** or an alternate name, that you can use in the rest of the statement. You create an alias by typing the name of the table, pressing the Spacebar, and then typing the name of the alias. No commas or periods are necessary to separate the two names.

You can use an alias for two basic reasons. The first reason is for simplicity. In Example 8, you will assign the SALES_REP table the alias S and the CUSTOMER table the alias C. By doing this, you can type S instead of SALES_REP and C instead of CUSTOMER in the remainder of the query. The query in this example is relatively simple, so you might not see the full benefit of this feature. If the query is complex and requires you to qualify the names, using aliases can simplify the process greatly.

EXAMPLE 8   List the sales rep number, last name, and first name for every sales rep together with the customer number, last name, and first name for each customer the sales rep represents.

The query format and results using aliases appear in Figure 4.11.

FIGURE 4.11   Using aliases

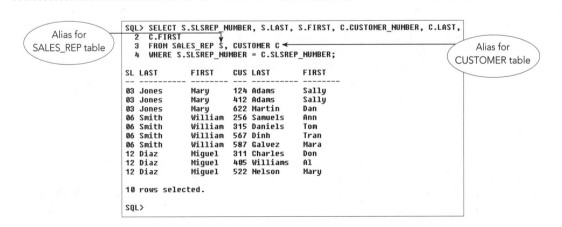

```
SQL> SELECT S.SLSREP_NUMBER, S.LAST, S.FIRST, C.CUSTOMER_NUMBER, C.LAST,
  2   C.FIRST
  3   FROM SALES_REP S, CUSTOMER C
  4   WHERE S.SLSREP_NUMBER = C.SLSREP_NUMBER;

SL LAST        FIRST    CUS LAST        FIRST
-- ----------  -------  --- ----------  --------
03 Jones       Mary     124 Adams       Sally
03 Jones       Mary     412 Adams       Sally
03 Jones       Mary     622 Martin      Dan
06 Smith       William  256 Samuels     Ann
06 Smith       William  315 Daniels     Tom
06 Smith       William  567 Dinh        Tran
06 Smith       William  587 Galvez      Mara
12 Diaz        Miguel   311 Charles     Don
12 Diaz        Miguel   405 Williams    Al
12 Diaz        Miguel   522 Nelson      Mary

10 rows selected.

SQL>
```

Alias for SALES_REP table

Alias for CUSTOMER table

In addition to their role in simplifying queries, aliases are essential in certain situations, as illustrated in Example 9.

■ More Complex Joins

The joins you have seen so far have been simple. Sometimes, however, you need to join a table to itself or join more than one table in a query.

Joining a Table to Itself

Sometimes it is necessary to join a table to itself, as illustrated in Example 9.

EXAMPLE 9 : Find every pair of customers who have the same first and last names.

If you had two separate tables for customers and the query requested customers in the first table having the same name as customers in the second table, you could use a regular join operation to find the answer. Here, however, there is only one table (CUSTOMER) that stores all the customer names. You actually can treat the CUSTOMER table as two tables in the query by creating an alias as illustrated in Example 8. In this case, you would change the FROM clause to:

```
FROM CUSTOMER F, CUSTOMER S
```

SQL treats this clause as a query of two tables: one that has the alias F, and another that has the alias S. The fact that both tables are really the single CUSTOMER table is not a problem. The query format and results appear in Figure 4.12.

FIGURE 4.12   Using aliases to join a table to itself

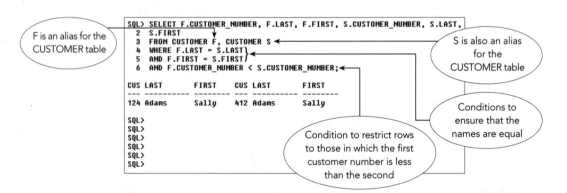

You are requesting a customer number, last name, and first name from the F table, followed by a customer number, last name, and first name from the S table. The query is subject to three conditions: the last names must match, the first names must match, and the customer number from the first table must be less than the customer number from the second table.

QUESTION   Why is the condition F.CUSTOMER_NUMBER < S.CUSTOMER_NUMBER important in the query formulation?

ANSWER   If you did not include this condition, the query result would be as shown in Figure 4.13. The first row is included because it is true that customer number 124 (Sally Adams) has the same name as customer number 124 (Sally Adams). The second row indicates that customer number 124 (Sally Adams) has the same name as customer number 412 (Sally Adams). This important information should be included. The next row, however, repeats the same information because customer number 412 (Sally Adams) has the same name as customer number 124 (Sally Adams). Of these three rows, the only row that should be included in the query results is the second row. The second row is also the only one of the three rows in which the first customer number (124) is less than the second customer number (412). This is why the actual query includes the condition F.CUSTOMER_NUMBER < S.CUSTOMER_NUMBER.

FIGURE 4.13

Incorrect joining
of a table to itself

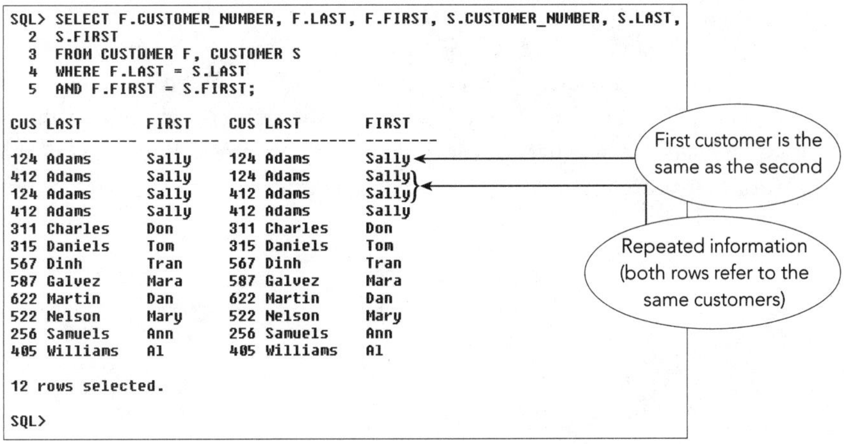

```
SQL> SELECT F.CUSTOMER_NUMBER, F.LAST, F.FIRST, S.CUSTOMER_NUMBER, S.LAST,
  2    S.FIRST
  3    FROM CUSTOMER F, CUSTOMER S
  4    WHERE F.LAST = S.LAST
  5    AND F.FIRST = S.FIRST;

CUS LAST        FIRST     CUS LAST        FIRST
--- ---------- -------- --- ---------- --------
124 Adams       Sally     124 Adams       Sally
412 Adams       Sally     124 Adams       Sally
124 Adams       Sally     412 Adams       Sally
412 Adams       Sally     412 Adams       Sally
311 Charles     Don       311 Charles     Don
315 Daniels     Tom       315 Daniels     Tom
567 Dinh        Tran      567 Dinh        Tran
587 Galvez      Mara      587 Galvez      Mara
622 Martin      Dan       622 Martin      Dan
522 Nelson      Mary      522 Nelson      Mary
256 Samuels     Ann       256 Samuels     Ann
405 Williams    Al        405 Williams    Al

12 rows selected.

SQL>
```

First customer is the same as the second

Repeated information (both rows refer to the same customers)

Joining Four Tables

It is possible to join several tables, as illustrated in Example 10. For each pair of tables you join, you include a condition indicating how the columns are related.

EXAMPLE 10 : For every part on order, list the part number, number ordered, order number, order date, customer number, last name, and first name of the customer who placed the order, and the last name and first name of the sales rep who represents each customer.

A part is on order if it occurs in any row in the ORDER_LINE table. The part number, number ordered, and order number are all found in the ORDER_LINE table. If these requirements were the only requirements of the query, the formulation would be as follows:

```
SELECT PART_NUMBER, NUMBER_ORDERED, ORDER_NUMBER
FROM ORDER_LINE
```

This formulation is not sufficient, however. You also need the order date and customer number, which are in the ORDERS table; the customer last name and first name, which are in the CUSTOMER table; and the sales rep last name and first name, which are in the SALES_REP table. Thus, you need to join *four* tables: ORDER_LINE, ORDERS, CUSTOMER, and SALES_REP. The procedure for joining four tables is essentially the same as the one for joining two tables. The difference is that the condition in the WHERE clause will be a compound condition. The WHERE clause in this case is written as follows:

```
WHERE ORDERS.ORDER_NUMBER = ORDER_LINE.ORDER_NUMBER
AND CUSTOMER.CUSTOMER_NUMBER = ORDERS.CUSTOMER_NUMBER
AND SALES_REP.SLSREP_NUMBER = CUSTOMER.SLSREP_NUMBER
```

The first condition relates an order line to an order with a matching order number. The second condition relates the order to the customer with a matching customer number. The final condition relates the customer to a sales rep based on a matching sales rep number.

For the complete query, you list all the desired columns in the SELECT clause and qualify any names that appear in more than one table. In the FROM clause, you list all four tables that are involved in the query. The query format and results appear in Figure 4.14.

FIGURE 4.14   Joining four tables

```
SQL> SELECT PART_NUMBER, NUMBER_ORDERED, ORDER_LINE.ORDER_NUMBER,          ( Tables to include
   2  ORDERS.CUSTOMER_NUMBER, CUSTOMER.LAST, CUSTOMER.FIRST,                (   in query
   3  SALES_REP.LAST, SALES_REP.FIRST
   4  FROM ORDER_LINE, ORDERS, CUSTOMER, SALES_REP
   5  WHERE ORDERS.ORDER_NUMBER = ORDER_LINE.ORDER_NUMBER            }       ( Conditions to relate
   6  AND CUSTOMER.CUSTOMER_NUMBER = ORDERS.CUSTOMER_NUMBER                  (   the tables
   7  AND SALES_REP.SLSREP_NUMBER = CUSTOMER.SLSREP_NUMBER;

PART NUMBER_ORDERED ORDER CUS LAST       FIRST     LAST       FIRST
---- -------------- ----- --- ---------- --------- ---------- --------
AX12             11 12489 124 Adams      Sally     Jones      Mary
BT04              1 12500 124 Adams      Sally     Jones      Mary
CB03              4 12494 315 Daniels    Tom       Smith      William
CX11              2 12495 256 Samuels    Ann       Smith      William
BT04              1 12491 311 Charles    Don       Diaz       Miguel
CZ81              2 12504 522 Nelson     Mary      Diaz       Miguel
BZ66              1 12491 311 Charles    Don       Diaz       Miguel
AZ52              2 12498 522 Nelson     Mary      Diaz       Miguel
BA74              4 12498 522 Nelson     Mary      Diaz       Miguel

9 rows selected.

SQL>
```

QUESTION   Why don't you have to qualify the PART_NUMBER column, which also appears as a column in the PART table?

ANSWER   If the PART table were used as one of the tables in the query, you would have to qualify PART_NUMBER; because this is not the case, the qualification is unnecessary. Among the tables listed in the query, only one column is labeled PART_NUMBER.

Certainly this last query is more involved than many of the previous ones. You might think that SQL is not such an easy language to use after all. If you take it one step at a time, however, the query in Example 10 really isn't that difficult. To construct a detailed query in a step-by-step fashion, do the following:

1. List in the SELECT clause all the columns that you want to display. If the name of any column appears in more than one table, precede the column name with the table name (that is, qualify the column name).

2. List in the FROM clause all the table names involved in the query. Usually you include the table names that contain the columns listed in the SELECT clause. Occasionally, however, there might be a table that does not contain any columns used in the SELECT clause but that does contain columns used in the WHERE clause. In this case, you must also list the table name in the FROM clause. For example, if you do not need to list a customer number or name, but you need to list the sales rep name, you wouldn't need to list any columns from the CUSTOMER

table in the SELECT clause. The CUSTOMER table is still required, however, because you must use columns from it in the WHERE clause.

3. Take one pair of tables at a time, and indicate in the WHERE clause the condition that relates the tables. Join these conditions with the AND operator. If there are any other conditions, include them in the WHERE clause and connect them to the other conditions with the AND operator. For example, if you wanted parts present on orders placed by only those customers having a $1,000 credit limit, you would add one more condition to the WHERE clause, as shown in Figure 4.15.

```
SQL> SELECT PART_NUMBER, NUMBER_ORDERED, ORDER_LINE.ORDER_NUMBER,
  2  ORDERS.CUSTOMER_NUMBER, CUSTOMER.LAST, CUSTOMER.FIRST,
  3  SALES_REP.LAST, SALES_REP.FIRST
  4  FROM ORDER_LINE, ORDERS, CUSTOMER, SALES_REP
  5  WHERE ORDERS.ORDER_NUMBER = ORDER_LINE.ORDER_NUMBER
  6  AND CUSTOMER.CUSTOMER_NUMBER = ORDERS.CUSTOMER_NUMBER
  7  AND SALES_REP.SLSREP_NUMBER = CUSTOMER.SLSREP_NUMBER
  8  AND CREDIT_LIMIT = 1000;

PART NUMBER_ORDERED ORDER CUS LAST       FIRST    LAST       FIRST
---- --------------- ----- --- ---------- -------- ---------- --------
AX12             11 12489 124 Adams      Sally    Jones      Mary
BT04              1 12500 124 Adams      Sally    Jones      Mary
BT04              1 12491 311 Charles    Don      Diaz       Miguel
B266              1 12491 311 Charles    Don      Diaz       Miguel

SQL>
```

■ Set Operations

In SQL, you can use the normal set operations: union, intersection, and difference. The **union** of two tables is a table containing every row that is in either the first table or the second table, or both. The **intersection** (**intersect**) of two tables is a table containing every row that is in both tables. The **difference** (**minus**) of two tables is the set of every row that is in the first table but not in the second table.

For example, suppose that TEMP1 is a table containing the customer number, last name, and first name of every customer represented by sales rep 12. Further suppose that TEMP2 is a table containing the customer number, last name, and first name of those customers who currently have orders on file (see Figure 4.16a).

The union of TEMP1 and TEMP2 (TEMP1 UNION TEMP2) consists of the customer number, last name, and first name of those customers who are represented by sales rep number 12 *or* who currently have orders on file, *or* both. The intersection of these two tables (TEMP1 INTERSECT TEMP2) contains those customers who are represented by sales rep number 12 *and* who have orders on file. The difference of these two tables (TEMP1 MINUS TEMP2) contains those customers who are represented by sales rep number 12 but *who do not* have orders on file. The results of these set operations are shown in Figure 4.16b.

TEMP1

CUSTOMER_NUMBER	LAST	FIRST
311	Charles	Don
405	Williams	Al
522	Nelson	Mary

TEMP2

CUSTOMER_NUMBER	LAST	FIRST
124	Adams	Sally
256	Samuels	Ann
311	Charles	Don
315	Daniels	Tom
522	Nelson	Mary

FIGURE 4.16B   Union, intersect, and minus of the TEMP1 and TEMP2 tables

TEMP1 UNION TEMP2

CUSTOMER_NUMBER	LAST	FIRST
124	Adams	Sally
256	Samuels	Ann
311	Charles	Don
315	Daniels	Tom
405	Williams	Al
522	Nelson	Mary

TEMP1 INTERSECT TEMP2

CUSTOMER_NUMBER	LAST	FIRST
311	Charles	Don
522	Nelson	Mary

TEMP1 MINUS TEMP2

CUSTOMER_NUMBER	LAST	FIRST
405	Williams	Al

There is an obvious restriction on the set operations. It does not make sense, for example, to talk about the union of the CUSTOMER table and the ORDERS table. What might rows in this union look like? The two tables in the union *must* have the same structure; that is, they must be union-compatible. Two tables are **union-compatible** if they have the same number of columns and if their corresponding columns have identical data types and lengths.

Note that the definition of union-compatible does not state that the columns of the two tables must be identical but rather that the columns must be of the same type. Thus, if one column is CHAR(20), the matching column also must be CHAR(20).

EXAMPLE 11   List the customer number, last name, and first name for every customer who is either represented by sales rep number 12 or who currently has orders on file, or both.

You can create a table containing the customer number, last name, and first name for every customer who is represented by sales rep number 12. You do this by selecting the customer numbers and names from the CUSTOMER table for which the sales rep number is 12. Then you can create another table containing the customer number, last name, and first name for every customer who currently has orders on file. You do this by creating a join of the CUSTOMER and ORDERS tables. The two tables created by this process have the same structure; that is, the following three columns: CUSTOMER_NUMBER, LAST, and FIRST. Because the tables are union-compatible, it is possible to take the union of these two tables. The union format and results appear in Figure 4.17.

FIGURE 4.17   Using the UNION operator

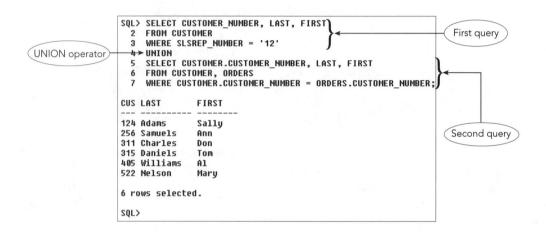

If the SQL implementation truly supports the union operation, it will remove any duplicate rows automatically. For example, any customer who is represented by sales rep number 12 *and* who currently has orders on file will not be listed twice. Oracle correctly removes duplicates. Some implementations of SQL, however, support a union operation but do not remove such duplicates.

EXAMPLE 12 : List the customer number, last name, and first name for every customer who is represented by sales rep number 12 and who currently has orders on file.

The only difference between this query and the one in Example 11 is that here the appropriate operator is INTERSECT. The query and its results appear in Figure 4.18.

FIGURE 4.18   Using the INTERSECT operator

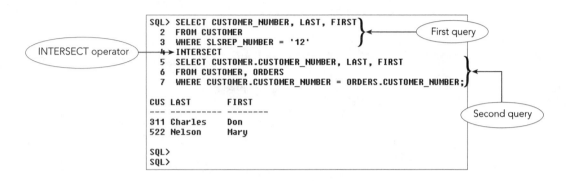

EXAMPLE 13   List the customer number, last name, and first name for every customer who is either represented by sales rep number 12 or who does not have orders currently on file.

The only difference between this query and the ones in Examples 11 and 12 is that here the appropriate operator is MINUS. The query and its results appear in Figure 4.19.

FIGURE 4.19   Using the MINUS operator

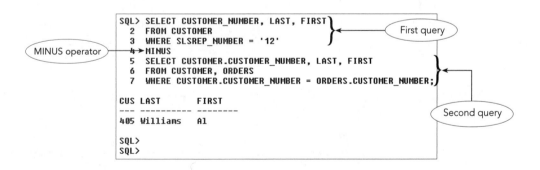

■ ALL and ANY

You can use the ALL and ANY operators with subqueries to produce a single column of numbers. If you precede the subquery by the ALL operator, the condition is true only if it satisfies *all* values produced by the subquery. If you precede the subquery by the ANY operator, the condition is true if it satisfies *any* value (one or more) produced by the subquery. The next examples illustrate the use of these operators.

EXAMPLE 14   Find the customer number, last name, first name, current balance, and sales rep number for every customer whose balance is greater than the individual balances of every customer of sales rep 12.

This query can be satisfied by finding the maximum balance of the customers represented by sales rep number 12 in a subquery and then finding all customers whose balance is greater than this number; however, there is an alternative. You can use the ALL operator, as shown in Figure 4.20, to simplify the process.

FIGURE 4.20

SELECT
command with
an ALL condition

```
SQL> SELECT CUSTOMER_NUMBER, LAST, FIRST, BALANCE, SLSREP_NUMBER
  2   FROM CUSTOMER
  3   WHERE BALANCE > ALL
  4   (SELECT BALANCE
  5   FROM CUSTOMER
  6   WHERE SLSREP_NUMBER = '12');

CUS LAST        FIRST      BALANCE SL
--- ---------- --------- --------- --
412 Adams       Sally       1817.5 03
622 Martin      Dan        1045.75 03

SQL>
SQL>
```

To some users, this formulation might seem more natural than finding the maximum balance in the subquery. For other users, the opposite might be true. You can employ whichever approach you prefer.

QUESTION   How would you get the same result for Example 14 without using the ALL operator?

ANSWER   You could query for every customer whose balance is greater than the maximum balance of any customer of sales rep number 12, as shown in Figure 4.21.

FIGURE 4.21

Alternative to ALL condition

```
SQL> SELECT CUSTOMER_NUMBER, LAST, FIRST, BALANCE, SLSREP_NUMBER
  2  FROM CUSTOMER
  3  WHERE BALANCE >
  4  (SELECT MAX(BALANCE)
  5  FROM CUSTOMER
  6  WHERE SLSREP_NUMBER = '12');

CUS LAST       FIRST      BALANCE SL
--- ---------- -------- --------- --
412 Adams      Sally       1817.5 03
622 Martin     Dan        1045.75 03

SQL>
SQL>
```

EXAMPLE 15 Find the customer number, last name, first name, current balance, and sales rep number of every customer whose balance is larger than the balance of at least one customer of sales rep number 12.

This query can be satisfied by finding the minimum balance of the customers represented by sales rep number 12 in a subquery and then finding all customers whose balance is greater than this number. Again there is an alternative. You can use the ANY operator, as shown in Figure 4.22, to simplify the process.

FIGURE 4.22

SELECT command with an ANY condition

```
SQL> SELECT CUSTOMER_NUMBER, LAST, FIRST, BALANCE, SLSREP_NUMBER
  2  FROM CUSTOMER
  3  WHERE BALANCE > ANY
  4  (SELECT BALANCE
  5  FROM CUSTOMER
  6  WHERE SLSREP_NUMBER = '12');

CUS LAST       FIRST      BALANCE SL
--- ---------- -------- --------- --
124 Adams      Sally       818.75 03
311 Charles    Don         825.75 12
315 Daniels    Tom         770.75 06
405 Williams   Al          402.75 12
412 Adams      Sally       1817.5 03
567 Dinh       Tran         402.4 06
587 Galvez     Mara         114.6 06
622 Martin     Dan        1045.75 03

8 rows selected.

SQL>
```

How would you get the same results without using the ANY operator?

You could query for the customers whose balance is greater than the minimum balance of any customers of sales rep number 12, as shown in Figure 4.23.

FIGURE 4.23

Alternative to
ANY condition

```
SQL> SELECT CUSTOMER_NUMBER, LAST, FIRST, BALANCE, SLSREP_NUMBER
  2  FROM CUSTOMER
  3  WHERE BALANCE >
  4  (SELECT MIN(BALANCE)
  5  FROM CUSTOMER
  6  WHERE SLSREP_NUMBER = '12');

CUS LAST         FIRST      BALANCE SL
--- ----------   --------   --------- --
124 Adams        Sally       818.75 03
311 Charles      Don         825.75 12
315 Daniels      Tom         770.75 06
405 Williams     Al          402.75 12
412 Adams        Sally      1817.5  03
567 Dinh         Tran        402.4  06
587 Galvez       Mara        114.6  06
622 Martin       Dan        1045.75 03

8 rows selected.

SQL>
SQL>
```

In this chapter, you learned how to use SQL commands to join tables. You qualified column names when necessary to obtain the desired results. You used subqueries to query multiple tables and used the IN and EXISTS operators in the process. You used an alias and saw situations where using aliases is beneficial. You used the SQL set operators UNION, INTERSECT, and MINUS to find rows in one, two, or both tables. Finally, you used the ALL and ANY operators to simplify certain SQL queries. In the next chapter, you will use SQL commands to update the data in your tables.

◾SUMMARY

1. To join tables together, indicate in the SELECT clause all columns to display, list in the FROM clause all tables to join, and then include in the WHERE clause any conditions requiring values in matching columns to be equal.

2. When referring to matching columns in different tables, you must qualify the column names to avoid confusion. You qualify column names using the following format: table name.column name.

3. Use the IN operator or the EXISTS command with an appropriate subquery as an alternate way of performing a join.

4. A subquery can contain another subquery. The innermost subquery is executed first.

5. The name of a table in a FROM clause can be followed by an alias, which is an alternate name for the table. The alias can be used in place of the table name throughout the SQL command.

6. By using two different aliases for the same table in a single SQL command, you can join a table to itself.

7. The UNION command creates a union of two tables; that is, the collection of rows that are in either or both tables. The INTERSECT command creates the intersection of two tables; that is, the collection of rows that are in both tables. The MINUS command creates the difference of two tables; that is, the collection of rows that are in the first table but not in the second table. To perform any of these operations, the tables must be union-compatible.

8. Two tables are union-compatible if they have the same number of columns and if their corresponding columns have identical data types and lengths.

9. If a subquery is preceded by the ALL command, the condition is true only if it is satisfied by *all* values produced by the subquery.

10. If a subquery is preceded by the ANY command, the condition is true if it is satisfied by *any* value (one or more) produced by the subquery.

◾EXERCISES (Premiere Products)

Use SQL to complete the following exercises.

Note: If you are using Oracle for these exercises and wish to print a copy of your commands and results, type SPOOL followed by the name of a file and then press the Enter key. All the commands from that point on will be saved in the file that you named. For example, to save the commands and results to a file named CHAPTER4.SQL on drive A, type the following command before beginning your work:

```
SPOOL A:CHAPTER4.SQL
```

When you have finished, type SPOOL OFF, and then press the Enter key to stop saving commands to the file. Then start any program that opens text files, open the file that you saved, and print it using the Print command on the File menu.

1. For every order, list the order number and order date along with the customer number, last name, and first name of the customer who placed the order.

2. For every order placed on September 5, 2002, list the order number and order date along with the customer number, last name, and first name of the customer who placed the order.

3. For every order, list the order number, order date, part number, number of units ordered, and quoted price for each order line that makes up the order.

4. Use the IN operator to find the customer number, last name, and first name for every customer who placed an order on September 5, 2002.

5. Repeat Exercise 4, but this time use the EXISTS operator in your answer.

6. Find the customer number, last name, and first name for every customer who did not place an order on September 5, 2002.

7. For every order, list the order number, order date, part number, part description, and item class for each part that makes up the order.

8. Repeat Exercise 7, but this time order the rows by item class and then by order number.

9. Use a subquery to find the sales rep number, last name, and first name for every sales rep who represents at least one customer with a credit limit of $2,000.

10. Repeat Exercise 9, but this time do not use a subquery.

11. Find the customer number, last name, and first name for every customer who currently has an order on file for an iron.

12. List the part number, part description, and item class for every pair of parts that are in the same item class and have the same warehouse number.

13. List the part description, part number, order number, and order date for every order placed by Mary Nelson that contains an order line for a treadmill.

14. List the part description, part number, order number, and order date for every order placed by Mary Nelson that does not contain an order line for a treadmill.

15. List the order number and order date for every order that was either placed by Mary Nelson or that contains an order line for an iron.

16. List the order number and order date for every order that was placed by Mary Nelson and that contains an order line for an iron.

17. List the order number and order date for every order that was placed by Mary Nelson but does not contain an order line for an iron.

18. List the part number, part description, unit price, and item class for every part that has a unit price greater than the unit price of every part in item class HW. Use

either the ALL or ANY operator in your query. (*Hint:* Be careful about which operator you use.)

19. If you used ALL in Exercise 18, repeat the exercise using ANY. If you used ANY, repeat the exercise using ALL. Then run the new command. What question does this command answer?

EXERCISES (Henry Books)

Use SQL to complete the following exercises.

1. For every book, list the book code, book title, publisher code, and publisher name.

2. For every book published by Signet, list the book title and book price.

3. List the book title and book code for every book published by Bantam Books that has a book price greater than $10.

4. List the book code, book title, and units on hand for every book in branch number 3.

5. List the book title for every book of type CS that is published by Best and Furrow.

6. Find the book title for every book written by author number 01. Use the IN operator in your formulation.

7. Repeat Exercise 6, but this time use the EXISTS operator in your formulation.

8. Find the book title and book code for every book located in branch number 2.

9. List every pair of book codes that have the same branch number in the INVENT table.

10. Find the units on hand, book title, and author's last name for every book in branch number 4.

11. Repeat Exercise 10, but this time list only paperback books.

12. Find the book code and title for every book whose price is over $5 or that was published in New York city.

13. Find the book code and title for every book whose price is over $5 and that was published in New York city.

14. Find the book code and title for every book whose price is over $5 but that was not published in New York city.

15. Find the book title and publisher code for every book whose price is greater than the book price of every book of type HOR.

16. Find the book title and publisher code for every book whose price is greater than the price of at least one book of type HOR.

CHAPTER 5

Updating Data

OBJECTIVES

- Use the COMMIT and ROLLBACK commands to make permanent data updates or to cancel updates

- Change data using the UPDATE command

- Add new data using the INSERT command

- Delete data using the DELETE command

- Create a new table from an existing table

- Use nulls in UPDATE commands

- Alter the rows in an existing table

- Change the structure of an existing table

Introduction

In this chapter, you will learn how to make changes to the data in a table. You will use the UPDATE command to change data in one or more rows in a table, use the INSERT command to add new rows, and use the DELETE command to delete rows. You will also create a new table from an existing table and use nulls in update operations. Finally, you will learn how to change the structure of a table in a variety of ways.

■ Commit and Rollback

When you update the data in a table, your updates are only temporary and you can cancel them at any time during your current work session. Updates become permanent automatically when you exit from the DBMS. During your current work session, however, you can make permanent changes immediately by running the COMMIT command. To make changes permanent during a work session, first execute the UPDATE command, and then type the COMMIT command, followed by a semicolon.

To cancel updates, use the ROLLBACK command. Any updates made since you ran the most recent COMMIT command will be reversed when you run the ROLLBACK command. If you have not run the COMMIT command, all updates made during the current session will be reversed. You should note that the ROLLBACK command only reverses changes to the data; it does not reverse changes made to a table's structure. For example, if you change the length of a character column, you cannot use the ROLLBACK command to return the column length to its original state.

■ Changing Existing Data in a Table

The data stored in your tables is subject to constant change—prices, addresses, commission amounts, and other data in a database change on a regular basis. To keep data current, you must be able to make these changes to the data in your tables. You can use the UPDATE command to change rows on which a specific condition is true.

EXAMPLE 1   Change the last name of customer number 256 in the Premiere Products database to Jones.

100

Use the SQL UPDATE command to make changes to existing data in the table. The format for using the UPDATE command is UPDATE <table name> SET <column name> = <new value>. When necessary, include a WHERE clause to indicate the row(s) in which the change is to take place. The command shown in Figure 5.1 changes the last name of customer number 256 to Jones. You know that the data has been updated when you see the confirmation message that one row has been updated. The SELECT command that follows the UPDATE command is an optional part of the update. The SELECT command shown in Figure 5.1 shows the data in the table after the change has been made. In general, you should always use a SELECT command to display the data you changed so you can verify that the change was made as you intended.

FIGURE 5.1   Updating a table

```
                         ┌─── Table to update ───┐
SQL> UPDATE CUSTOMER ◄───┘
  2    SET LAST = 'Jones' ◄──────────────────────── New value
  3    WHERE CUSTOMER_NUMBER = '256'; ◄──── Condition
1 row updated.    Column to change

SQL> SELECT *
  2    FROM CUSTOMER;

CUS LAST         FIRST     STREET           CITY             ST ZIP_C  BALANCE CREDIT_LIMIT SL
--- ----------   --------  ---------------  ---------------- -- -----  ------- ------------ --
124 Adams        Sally     481 Oak          Lansing          MI 49224  818.75         1000 03
256 Jones◄       Ann       215 Pete         Grant            MI 49219    21.5         1500 06
311 Charles      Don       48 College       Ira              MI 49034  825.75         1000 12
315 Daniels      Tom       914 Cherry       Kent             MI 48391  770.75          750 06
405 Williams     Al        519 Watson       Grant            MI 49219  402.75         1500 12
412 Adams        Sally     16 Elm           Lansing          MI 49224  1817.5         2000 03
522 Nelson       Mary      108 Pine         Ada              MI 49441   98.75         1500 12
567 Dinh         Tran      808 Ridge        Harper           MI 48421   402.4          750 06
587 Galvez       Mara      512 Pine         Ada              MI 49441   114.6         1000 06
622 Martin       Dan       419 Chip         Grant            MI 49219 1045.75         1000 03

10 rows selected.      Updated value
                        in table
SQL>
```

If you determine that the update was made incorrectly, you can use the ROLLBACK command to return the data to its original state. As shown in Figure 5.2, the name of customer number 256 has been returned to its original value (Samuels).

FIGURE 5.2   Using the ROLLBACK command

```
SQL> ROLLBACK;

Rollback complete.

SQL> SELECT *
  2  FROM CUSTOMER;

CUS LAST        FIRST    STREET       CITY            ST ZIP_C   BALANCE CREDIT_LIMIT SL
--- ----------  -------- -----------  --------------- -- -----  ------- ------------ --
124 Adams       Sally    481 Oak      Lansing         MI 49224  818.75          1000 03
256 Samuels     Ann      215 Pete     Grant           MI 49219    21.5          1500 06
311 Charles     Don      48 College   Ira             MI 49034  825.75          1000 12
315 Daniels     Tom      914 Cherry   Kent            MI 48391  770.75           750 06
405 Williams    Al       519 Watson   Grant           MI 49219  402.75          1500 12
412 Adams       Sally    16 Elm       Lansing         MI 49224  1817.5          2000 03
522 Nelson      Mary     108 Pine     Ada             MI 49441   98.75          1500 12
567 Dinh        Tran     808 Ridge    Harper          MI 48421   402.4           750 06
587 Galvez      Mara     512 Pine     Ada             MI 49441   114.6          1000 06
622 Martin      Dan      419 Chip     Grant           MI 49219 1045.75          1000 03

10 rows selected.

SQL>
```

ROLLBACK command

Change is reversed

If, on the other hand, you have verified that the updates you made are correct, you can use the COMMIT command to make the update permanent. You do this by typing COMMIT; after running the update. However, you should note that the COMMIT command is permanent; running the ROLLBACK command cannot reverse the update.

Note: Normally you would use the ROLLBACK command only to correct a problem with an update. If you execute the commands in the body of this chapter, however, you should execute a rollback after each UPDATE, INSERT, or DELETE command to return your data to its original state before you complete the exercises at the end of the chapter. You will also need to execute rollbacks when you complete the end-of-chapter exercises.

EXAMPLE 2:   For each customer with a $1,000 credit limit whose balance does not exceed the credit limit, increase the credit limit to $1,200.

The only difference between Examples 1 and 2 is that Example 2 uses a compound condition to identify the rows to change. The UPDATE command and the SELECT command that shows its results appear in Figure 5.3.

FIGURE 5.3   Using a compound condition to update a table

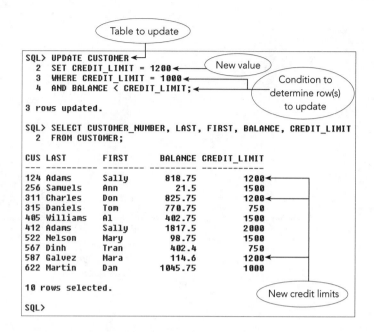

You also can use the previous value in a column in the update. For example, if you need to increase the credit limit by 10% instead of changing it to a specific value, you can multiply the previous credit limit by 1.10. The SET clause to do this is as follows:

```
SET CREDIT_LIMIT = CREDIT_LIMIT * 1.10
```

▪ Adding New Rows To An Existing Table

In Chapter 2, you used the INSERT command to add data to the database. You can also use the INSERT command as a way of updating the data in a table.

EXAMPLE 3 : Add sales rep number 14 to the SALES_REP table. Her name is Ann Crane, her address is 123 River in Alpen, MI, and her zip code is 49114. Her commission rate is 5% (0.05), but she has not yet earned any commission.

You can add new data by using the INSERT command, as shown in Figure 5.4.

FIGURE 5.4   Inserting a row

```
                New row added                Table to update

      SQL> INSERT INTO SALES_REP
         2    VALUES
         3    ('14','Crane','Ann','123 River','Alpen','MI','49114',0,0.05);      Values for new row

      1 row created.

      SQL> SELECT *
         2    FROM SALES_REP;

      SL LAST        FIRST     STREET           CITY         ST ZIP_C TOTAL_COMMISSION COMMISSION_RATE
      -- ---------   --------  ---------------  -----------  -- ----- ---------------- ---------------
      03 Jones       Mary      123 Main         Grant        MI 49219             2150             .05
      06 Smith       William   102 Raymond      Ada          MI 49441           4912.5             .07
      12 Diaz        Miguel    419 Harper       Lansing      MI 49224             2150             .05
      14 Crane       Ann       123 River        Alpen        MI 49114                0             .05

      SQL>
```

■ Deleting Existing Rows From A Table

When rows are no longer needed, you should remove them. For example, if Al Williams has moved and is not a customer of Premiere Products any longer, you should remove his row from the CUSTOMER table.

EXAMPLE 4   :   Delete from the database the customer information for Al Williams.

To delete data from the database, use the DELETE command. The format for the DELETE command is DELETE <table name> WHERE <column name> = <value>. The command shown in Figure 5.5 deletes from the CUSTOMER table any row with a value of "Williams" in the LAST column.

FIGURE 5.5   Deleting a row

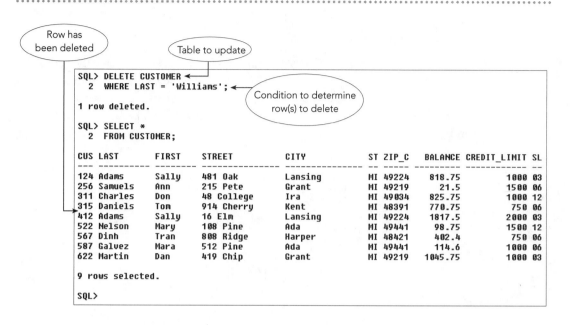

Deleting rows can be dangerous. If other customers in the table also have the last name of Williams, those customers' records will also be deleted by the DELETE command shown in Figure 5.5. The safest way to delete data is to use a condition that involves the value in the primary key. Because the primary key value, by definition, is a unique value, you can be certain that you will not delete any other rows in the table accidentally.

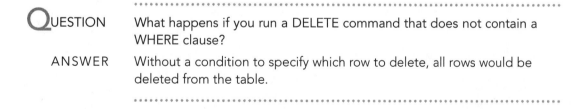

QUESTION   What happens if you run a DELETE command that does not contain a WHERE clause?

ANSWER   Without a condition to specify which row to delete, all rows would be deleted from the table.

■ Creating a New Table From an Existing Table

It is possible to create a new table by using some of the data in an existing table, as illustrated in the following two examples.

EXAMPLE 5 Create a new table named SMALL_CUST containing the same columns as the CUSTOMER table, and then insert into the new table only those rows for which the credit limit is $1,200 or less.

The first task is to describe the new table named SMALL_CUST by using the CREATE TABLE command, as shown in Figure 5.6. Then you can use the INSERT command to add data to the table. Here, however, you can use a SELECT command to indicate which rows from the existing table (CUSTOMER) to insert into the new table. The INSERT command appears in Figure 5.6.

FIGURE 5.6

Creating a table from an existing table

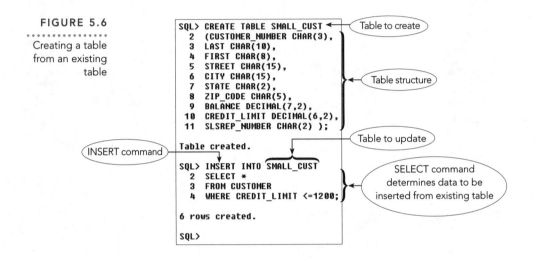

The SELECT command shown in Figure 5.7 yields the resulting data in the new SMALL_CUST table.

FIGURE 5.7 Results of table creation

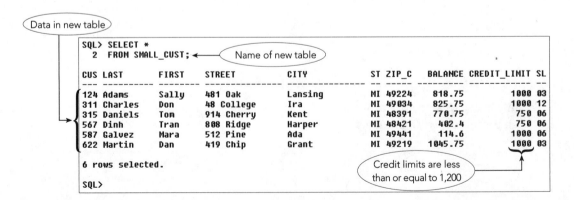

106

Note: In Example 2, an UPDATE command was used to change certain credit limits to $1,200. Those changes are not reflected in the data shown in Figure 5.7 because a rollback was executed after the update. If you are executing the updates in the body of the chapter, but you are not executing a rollback after each one, your results will be different from those shown in Figure 5.7.

When you create a table from an existing table, you do not have to include the same columns as in the existing table, as shown in the next example.

EXAMPLE 6   Create a new table named CUST_OF_03, containing the customer number, last name, first name, and balance of every customer assigned to sales rep number 03.

The CREATE TABLE command for this new table appears in Figure 5.8. The INSERT command to add the data from the existing CUSTOMER table and the resulting data in the new table appears in Figure 5.9.

FIGURE 5.8
. . . .
Creating a table
that does not
contain all the
columns

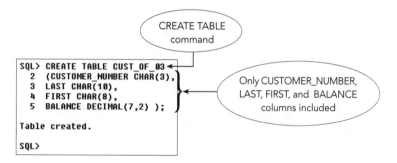

FIGURE 5.9
. . . .
Inserting rows

■ Changing a Value in a Column to Null

There are some special issues involved when dealing with nulls. You already have seen how to add a row in which some of the values are null and how to select rows in which a given column is null. You must also be able to change the value in a column in an existing row to null, as shown in Example 7. Remember that in order to make this type of change, the affected column must be able to accept nulls. If you specified NOT NULL for the column when you created the table, then changing a value in a column to null is prohibited.

EXAMPLE 7 Change the address of customer number 124 in the CUSTOMER table to null.

The command for changing the value to null is exactly what it would be for changing any other value. You simply use the value NULL as the replacement value, as shown in Figure 5.10. Notice that the NULL command is *not* enclosed in single quotation marks. If it were, the command would change the street address to the word NULL.

FIGURE 5.10   Setting a column to null

```
SQL> UPDATE CUSTOMER ←                              Table to update
  2   SET STREET = NULL ←
  3   WHERE CUSTOMER_NUMBER = 124;        Change value
                                            to null
1 row updated.           Column to
                          update
SQL> SELECT *
  2   FROM CUSTOMER;              Value is null

CUS LAST       FIRST    STREET        CITY          ST ZIP_C   BALANCE CREDIT_LIMIT SL
--- ---------- -------- ------------- ------------- -- ----- --------- ------------- --
124 Adams      Sally                  Lansing       MI 49224   818.75          1000 03
256 Samuels    Ann      215 Pete      Grant         MI 49219    21.5           1500 06
311 Charles    Don      48 College    Ira           MI 49034   825.75          1000 12
315 Daniels    Tom      914 Cherry    Kent          MI 48391   770.75           750 06
405 Williams   Al       519 Watson    Grant         MI 49219   402.75          1500 12
412 Adams      Sally    16 Elm        Lansing       MI 49224  1817.5           2000 03
522 Nelson     Mary     108 Pine      Ada           MI 49441    98.75          1500 12
567 Dinh       Tran     808 Ridge     Harper        MI 48421   402.4            750 06
587 Galvez     Mara     512 Pine      Ada           MI 49441   114.6           1000 06
622 Martin     Dan      419 Chip      Grant         MI 49219  1045.75          1000 03

10 rows selected.

SQL>
```

■ Changing a Table's Structure

One of the nicest features of a relational DBMS is the ease with which you can change a table's structure. You can use the CREATE TABLE command to add new tables, delete tables that are no longer required, add new columns to a table, and change the physical characteristics of existing columns. Next, you will see how to accomplish these changes.

Altering the Structure of an Existing Table

Using SQL to alter the structure of an existing table is easy. In contrast, altering the structure of an existing table in a nonrelational system is a much more complex process. In a nonrelational database system you have to change the description of the structure by using utility programs that unload the data from the current structure and then reload it with the new structure.

With SQL, you can change a table's structure by using the ALTER TABLE command, as illustrated in the following examples.

EXAMPLE 8 ⋮ Premiere Products decides to maintain a customer type for each customer in the database. These types are R for regular customers, D for distributors, and S for special customers. Add this information as a new column in the CUSTOMER table.

To add a new column, use the ADD clause of the ALTER TABLE command. The format for the command is ALTER TABLE <table name> ADD <column name> <characteristics>. Figure 5.11 shows the appropriate ALTER TABLE command for this example.

FIGURE 5.11   Adding a column using ALTER

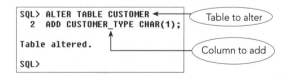

The CUSTOMER table now contains a new column named CUSTOMER_TYPE. Any new rows added to the table must include values for the new column. Effective immediately, all existing rows contain this new column. The data in any existing row will be changed to reflect the new column the next time the row is updated. However, any time a row is selected for any reason, the system treats the row as though the column is actually present. Thus, to the user, it will feel as though the structure was changed immediately.

For existing rows, some value of CUSTOMER_TYPE must be assigned. The simplest approach (from the point of view of the DBMS, not the user) is to assign the value NULL as a CUSTOMER_TYPE in all existing rows. This process requires the CUSTOMER_TYPE column to accept null values, and some systems actually insist on this. That is, any column added to a table definition *must* accept nulls; the user has no choice. A more flexible approach, and one that is supported by some systems, is to allow the user to specify an initial value. For example, if most customers are of type R, you might set all the types for existing customers to R, and then later change distributors to type D and special customers to type S. The command to change the structure and set the value in the CUSTOMER_TYPE column to R for all existing records is as follows:

```
ALTER TABLE CUSTOMER
ADD CUSTOMER_TYPE CHAR(1) INIT = 'R';
```

If a system will only set new columns to null, as is the case in Oracle, you can still accomplish the previous initialization by following the ALTER TABLE command with an UPDATE command, as shown in Figure 5.12.

FIGURE 5.12 Making the same entry in all rows

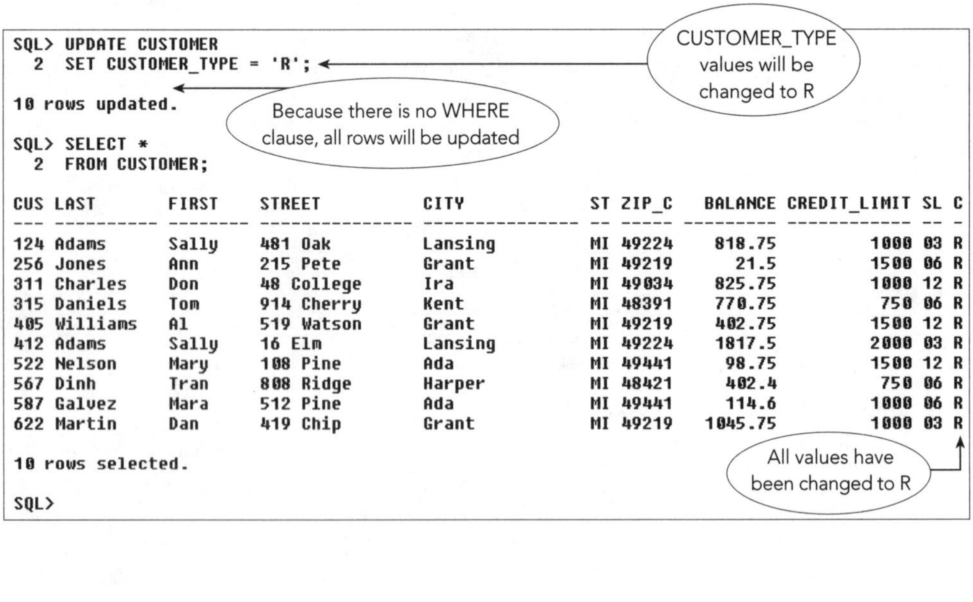

```
SQL> UPDATE CUSTOMER
  2    SET CUSTOMER_TYPE = 'R';  ←
                                        CUSTOMER_TYPE
                                        values will be
10 rows updated.                        changed to R
                        Because there is no WHERE
SQL> SELECT *           clause, all rows will be updated
  2    FROM CUSTOMER;

CUS LAST        FIRST    STREET          CITY          ST ZIP_C   BALANCE CREDIT_LIMIT SL C
--- ---------- -------- --------------- ------------- -- ----- --------- ------------ -- -
124 Adams       Sally    481 Oak         Lansing       MI 49224   818.75         1000 03 R
256 Jones       Ann      215 Pete        Grant         MI 49219    21.5          1500 06 R
311 Charles     Don      48 College      Ira           MI 49034   825.75         1000 12 R
315 Daniels     Tom      914 Cherry      Kent          MI 48391   770.75          750 06 R
405 Williams    Al       519 Watson      Grant         MI 49219   402.75         1500 12 R
412 Adams       Sally    16 Elm          Lansing       MI 49224  1817.5          2000 03 R
522 Nelson      Mary     108 Pine        Ada           MI 49441    98.75         1500 12 R
567 Dinh        Tran     808 Ridge       Harper        MI 48421   402.4           750 06 R
587 Galvez      Mara     512 Pine        Ada           MI 49441   114.6          1000 06 R
622 Martin      Dan      419 Chip        Grant         MI 49219  1045.75         1000 03 R

10 rows selected.
                                        All values have
SQL>                                    been changed to R
```

Two customers have a type other than R. The type for customer number 412 should be S, and the type for customer number 622 should be D.

The previous example assigned type R to every customer. To change individual types to something other than type R, use the UPDATE command. The appropriate UPDATE commands to make these changes appear in Figure 5.13.

FIGURE 5.13

Changing individual rows

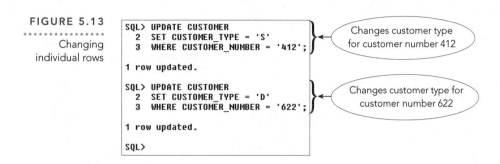

```
SQL> UPDATE CUSTOMER
  2    SET CUSTOMER_TYPE = 'S'
  3    WHERE CUSTOMER_NUMBER = '412';

1 row updated.

SQL> UPDATE CUSTOMER
  2    SET CUSTOMER_TYPE = 'D'
  3    WHERE CUSTOMER_NUMBER = '622';

1 row updated.

SQL>
```

Changes customer type for customer number 412

Changes customer type for customer number 622

Figure 5.14 shows the results of these UPDATE commands. The customer type for customer number 412 is S and the type for customer number 622 is D. The type for all other customers is R.

FIGURE 5.14 Results of changes

```
SQL> SELECT *
  2  FROM CUSTOMER;

CUS LAST        FIRST     STREET          CITY             ST ZIP_C    BALANCE CREDIT_LIMIT SL C
--- ----------  --------  --------------- ---------------- -- -----  --------- ------------ -- -
124 Adams       Sally     481 Oak         Lansing          MI 49224    818.75         1000 03 R
256 Jones       Ann       215 Pete        Grant            MI 49219     21.5          1500 06 R
311 Charles     Don       48 College      Ira              MI 49034    825.75         1000 12 R
315 Daniels     Tom       914 Cherry      Kent             MI 48391    770.75          750 06 R
405 Williams    Al        519 Watson      Grant            MI 49219    402.75         1500 12 R
412 Adams       Sally     16 Elm          Lansing          MI 49224   1817.5          2000 03 S
522 Nelson      Mary      108 Pine        Ada              MI 49441     98.75         1500 12 R
567 Dinh        Tran      808 Ridge       Harper           MI 48421    402.4           750 06 R
587 Galvez      Mara      512 Pine        Ada              MI 49441    114.6          1000 06 R
622 Martin      Dan       419 Chip        Grant            MI 49219   1045.75         1000 03 D

10 rows selected.

SQL>
```

Values have been changed as indicated

EXAMPLE 10 : The length of the STREET column is too short. Increase its length to 20 characters.

You can change the characteristics of existing columns by using the MODIFY clause of the ALTER TABLE command. Figure 5.15 shows the ALTER TABLE command that changes the length of the STREET column from 15 to 20 characters. The DESCRIBE command in the figure shows the new structure of the CUSTOMER table.

FIGURE 5.15 Changing the column width

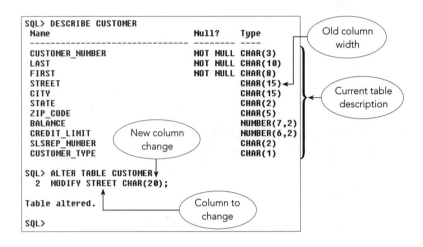

EXAMPLE 11 : Currently the CITY column can accept nulls. Change the CITY column so that nulls are not allowed. Currently the FIRST column is not allowed to accept nulls. Change the FIRST column so that nulls are allowed.

You can use the MODIFY clause of the ALTER TABLE command to change a column that currently allows nulls to NOT NULL. You also can change a column that doesn't allow nulls to NULL. Both changes appear in Figure 5.16. Now the DBMS will reject any attempt to store a null value in the CITY column, but it will accept a null value in the FIRST column.

FIGURE 5.16
· · · · · · · · · · · · · · · ·
Changing
columns to NOT
NULL or NULL

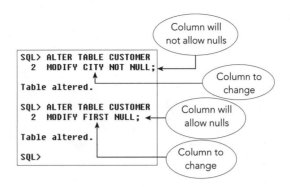

If there were existing rows in the CUSTOMER table in which the CITY column was already null, the DBMS would reject the modification to the CITY column shown in Figure 5.16. The DBMS would display an error message indicating that this alteration is not possible. In this case, you must first use an UPDATE command to change all values that are null to some other value. Then you could alter the table's structure as shown in Figure 5.16. Figure 5.17 shows the current description of the CUSTOMER table after making changes to its structure.

FIGURE 5.17   Current table structure

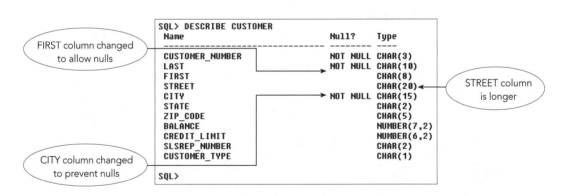

Making Complex Changes

In some cases, you might need to change a table's structure in ways that are beyond the capabilities of your DBMS. Perhaps you need to eliminate a column, change the column order, or combine data from two tables into one, but your system does not allow these types of changes. For example, some systems, including Oracle, do not allow you to reduce the size of a column or to change a data type. In these situations, you can use the CREATE

TABLE command to describe the new table, and then insert values into it using the INSERT command combined with an appropriate SELECT clause.

Dropping a Table

As you learned in Chapter 2, you can delete a table that is no longer needed by using the DROP command. For example, if the SALES_REP table is not needed in the Premiere Products database, you could delete it with the following command:

```
DROP TABLE SALES_REP;
```

If the command is executed, the SALES_REP table and all its data will be deleted permanently from the database.

In this chapter, you learned how to use the UPDATE command to change the data in a table. You learned to use the INSERT command to add new rows and used the DELETE command to delete existing rows. You created a new table that contained some of the data from an existing table. You used the ALTER TABLE command to change the structure of a table by adding new columns and changing the characteristics of existing columns. Finally, you learned how to use the DROP command to delete an entire table and its data. In the next chapter, you will use SQL commands for database administration tasks, such as creating views and using the system catalog.

◼ SUMMARY

1. Use the UPDATE command to change existing data in a table.

2. Use the COMMIT command to make updates permanent; use the ROLLBACK command to cancel any updates that have not been committed.

3. Use the INSERT command to add new rows to a table.

4. Use the DELETE command to delete existing rows from a table.

5. To create a new table from an existing table, first create the new table by using the CREATE TABLE command. Then use an INSERT command containing a SELECT clause to select the desired data to be included from the existing table.

6. To change all values in a column to null, the clause is SET <column name> = NULL. To change a specific value in a column to null, use a condition to select the row.

7. To add a column to a table, use the ALTER TABLE command with the ADD clause.

8. To change the characteristics of a column, use the ALTER TABLE command with the MODIFY clause.

9. Use the DROP TABLE command to delete a table and all of its data.

■EXERCISES (Premiere Products)

Use SQL to make the following changes to the Premiere Products database. After each change, execute an appropriate query to determine whether the change was made correctly. Remember to execute a rollback after each UPDATE, INSERT, or DELETE command.

Note: If you are using Oracle for these exercises and wish to print a copy of your commands and results, type SPOOL followed by the name of a file and then press the Enter key. All the commands from that point on will be saved in the file that you named. For example, to save the commands and results to a file named CHAPTER5.SQL on drive A, type the following command before beginning your work:

```
SPOOL A:CHAPTER5.SQL
```

When you have finished, type SPOOL OFF, and then press the Enter key to stop saving commands to the file. Then start any program that opens text files, open the file that you saved, and print it using the Print command on the File menu.

1. Change the part description of part number BT04 to Oven.

2. Add $100 to the credit limit of every customer represented by sales rep number 06.

3. Add order 12600 (order date: 9/06/2002, customer number: 311) to the database. Add two order lines for this order. On the first line, the part number is AX12, the number ordered is 5, and the quoted price is $13.95. On the second line, the part number is BA74, the number ordered is 3, and the quoted price is $4.50.

4. Write the command to delete every customer whose balance is $0 and who is represented by sales rep number 12. Do *not* execute the command.

5. Create a new table named SPGOOD to contain the columns PART_NUMBER, PART_DESCRIPTION, and UNIT_PRICE. Then insert into this new table the part number, part description, and unit price from the PART table for every part whose item class is SG. Do *not* execute a rollback.

6. Change the street address of sales rep number 03 in the SALES_REP table to null.

7. Add a column named ALLOCATION to the PART table. The allocation is a three-digit number representing the number of units of each part that have been allocated to each customer. Set all values of ALLOCATION to zero. Calculate the number of units of part number BT04 currently on order. Change the value of ALLOCATION for part number BT04 to this number.

8. Increase the length of the PART_DESCRIPTION column to 30 characters.

9. Write the command to remove the PART table from the Premiere Products database. Do *not* execute the command.

■EXERCISES (Henry Books)

Use SQL to make the following changes to the Henry Books database. After each change, execute an appropriate query to determine whether the change was made correctly. Remember to execute a rollback after each UPDATE, INSERT, or DELETE command.

1. Change the number of units on hand to 5 for all books located in branch number 1.

2. Bantam Books has decreased the price of its books by 3%. Update the prices in the BOOK table.

3. Insert a new book into the database. The book code is 9700, the title is Using Microsoft Access 2000, the publisher is Best and Furrow, the book type is computer science, the price is $19.97, and the book is available only in paperback. The author number is 07, and there are four books on hand in branch number 1. The sequence number is 2.

4. Write the command to delete the author named Robert Wray from the database. Make sure that your command will not delete any additional rows. Do *not* execute the command.

5. Create a new table named FICTION using the data in the book code, book title, and book price columns in the BOOK table. Select only those books of type FIC and insert them into the new table. Do *not* execute a rollback.

6. Change the book price of all books in the FICTION table by 12%. Do *not* execute a rollback.

7. The price of the book titled "Amerika" in the FICTION table has been increased to an unknown amount. Change the value to reflect this. Do *not* execute a rollback.

8. Add to the FICTION table a new character column that is one character in length named BEST_SELLER. The default value for all columns is N.

9. Change the BEST_SELLER column in the FICTION table to Y for the book titled "Kane and Abel."

10. Change the length of the BOOK_TITLE column in the FICTION table to 50.

11. Change the BEST_SELLER column in the FICTION table to reject nulls.

12. Change the BEST_SELLER column in the FICTION table to accept nulls.

13. Delete the FICTION table from the database.

CHAPTER **6**

Database Administration

OBJECTIVES

- Understand, create, and drop views

- Recognize the benefits of using views

- Grant and revoke users' database privileges

- Create, use, and drop an index

- Understand the purpose, advantages, and disadvantages of using an index

- Understand and obtain information from the system catalog

- Use integrity constraints to control data entry

Introduction

There are some special issues involved in managing a database. This process, often called **database administration**, is especially important when more than one person uses the database. In a business organization, a person or an entire group known as the **database administrator** is charged with managing the database.

In Chapter 5 you learned about one function of the database administrator: changing the structure of a database. In this chapter, you will learn about some additional tasks of the database administrator. You will see how the database administrator can give each user his or her own view of the database. You will use the GRANT and REVOKE commands to assign different database privileges to different users. You will use indexes to improve database performance. You will see how SQL keeps information about the database structure in a special object called the system catalog. Using the system catalog, the database administrator can obtain helpful information concerning the database structure, and specify integrity constraints to establish rules that the data in the database must satisfy.

■ Views

Most database management systems are capable of giving each user his or her own picture of the data in the database. In SQL this is done using views. The existing, permanent tables in a relational database are called **base tables**. A **view** is a derived table because the data in a view is derived from the base table. To the user, a view appears to be an actual table. In many cases, a user can interact with the database using a view. Because a view usually includes less information than the full database, using a view can represent a great simplification. Views also provide a measure of security, because omitting sensitive tables or columns from a view renders them unavailable to anyone who is accessing the database through the view.

To illustrate the idea of a view, suppose that Juan is interested in the part number, part description, units on hand, and unit price of parts in item class HW. He is not interested in any other columns in the PART table, nor is he interested in any of the rows that correspond to parts in other item classes. Whereas Juan cannot change the structure of the PART table and omit some of its rows for his own purposes, he can do the next best thing. He can create a view that consists of only the rows and columns that he needs.

A view is defined by creating a defining query. The **defining query** is a SQL command that indicates the rows and columns that will appear in the view. The command to create the view for Juan, including the defining query, is illustrated in Example 1.

EXAMPLE 1 Define a view named HOUSEWARES that consists of the part number, part description, units on hand, and unit price of all parts in item class HW.

The CREATE VIEW command shown in Figure 6.1 creates a view of the PART table that contains only the specified columns that match the selection condition.

FIGURE 6.1

Creating the
HOUSEWARES
view

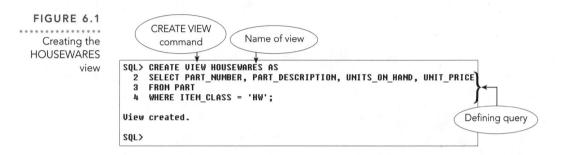

Given the current data in the Premiere Products database, the HOUSEWARES view contains the data shown in Figure 6.2.

FIGURE 6.2

HOUSEWARES
view

PART_NUMBER	PART_DESCRIPTION	UNITS_ON_HAND	UNIT_PRICE
AX12	Iron	104	$24.95
BH22	Cornpopper	95	$24.95
CA14	Griddle	78	$39.99
CX11	Blender	112	$22.95

Note: If you did not rollback the data in Chapter 5, your results will be different from those shown in Figure 6.2 and throughout Chapter 6.

The data does not actually exist in this form, however, nor will it *ever* exist in this form. When this view is used, it is tempting to think that the query will be executed and produce some sort of temporary table, named HOUSEWARES, that the user can access at any time. This is *not* what happens. Instead, the query acts as a sort of "window" into the database (see Figure 6.3). As far as a user of this view is concerned, the entire database consists of the dark shaded portion of the PART table.

FIGURE 6.3   Premiere Products sample data

PART

PART_NUMBER	PART_DESCRIPTION	UNITS_ON_HAND	ITEM_CLASS	WAREHOUSE_NUMBER	UNIT_PRICE
AX12	Iron	104	HW	3	$ 24.95
AZ52	Dartboard	20	SG	2	$ 12.95
BA74	Basketball	40	SG	1	$ 29.95
BH22	Cornpopper	95	HW	3	$ 24.95
BT04	Gas Grill	11	AP	2	$149.99
BZ66	Washer	52	AP	3	$399.99
CXA14	Griddle	78	HW	3	$ 39.99
CB03	Bike	44	SG	1	$299.99
CX11	Blender	112	HW	3	$ 22.95
CZ81	Treadmill	68	SG	2	$349.95

The way in which the view is implemented is clever. Suppose, for example, that a user of this view runs the query shown in Figure 6.4.

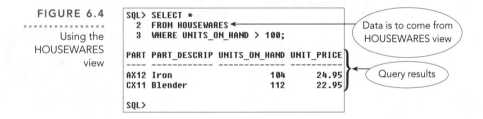

FIGURE 6.4

Using the HOUSEWARES view

Rather than being executed directly, the query first is merged with the query that defines the view, producing the following statement:

```
SELECT PART_NUMBER, PART_DESCRIPTION, UNITS_ON_HAND, UNIT_PRICE
FROM PART
WHERE ITEM_CLASS = 'HW'
AND UNITS_ON_HAND > 100;
```

Notice that the selection is from the PART table rather than from the HOUSEWARES view, the asterisk is replaced by just those columns in the HOUSEWARES view, and the condition includes the condition in the query entered by the user together with the condition stated in the view definition. This new query is the one that the DBMS actually executes.

The user, however, is unaware that this kind of activity takes place. There is no evidence of it shown in Figure 6.4. It seems as though there actually is a table named HOUSEWARES that is being accessed.

One advantage of this approach is that, because the HOUSEWARES view never exists in its own right, any update to the PART table is reflected *immediately* in the HOUSEWARES view and is apparent to anyone accessing the database through the view. If the HOUSEWARES view were an actual stored table, this immediate update would not be the case.

The formulation of a view definition is: CREATE <view name> AS <query>. The query, which is called the **defining query**, can be any valid SQL query.

You also can assign column names that are different from those in the base table, as illustrated in the next example.

EXAMPLE 2 : Define a view named HOUSEWARES that consists of the part number, part description, units on hand, and unit price of all parts in item class HW. In this view, rename the PART_NUMBER column to NUM, the PART_DESCRIPTION column to DSC, the UNITS_ON_HAND column to OH, and the UNIT_PRICE column to PRCE.

Note: If you created the HOUSEWARES view in Example 1, type DROP VIEW HOUSE-WARES; at the SQL prompt to delete it before executing the command shown in Figure 6.5.

When renaming columns, you include the new column names in parentheses following the name of the view, as shown in Figure 6.5. In this case, anyone accessing the HOUSEWARES view will refer to PART_NUMBER as NUM, to PART_DESCRIPTION as DSC, to UNITS_ON_HAND as OH, and to UNIT_PRICE as PRCE. If you select all columns from the HOUSEWARES view, the new column names will be displayed as shown in Figure 6.5.

FIGURE 6.5   Renaming columns when creating a view

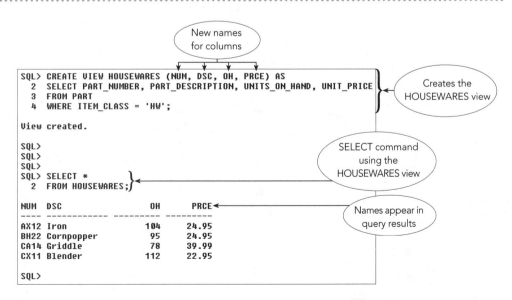

The HOUSEWARES view is an example of a **row-and-column subset view** because it consists of a subset of the rows and columns in some base table—in this case, in the PART table. Because the query can be any valid SQL query, a view could involve the join of two or more tables, or it also could involve statistics. The next example illustrates a view that joins two tables.

EXAMPLE 3   Define a view named SALES_CUST. The view consists of the sales rep number (named SNUMB), sales rep last name (named SLAST), sales rep first name (named SFIRST), customer number (named CNUMB), customer last name (named CLAST), and customer first name (named CFIRST) for all sales reps and matching customers in the SALES_REP and CUSTOMER tables.

The command to create this view appears in Figure 6.6.

FIGURE 6.6
···············
Creating the
SALES_CUST
view

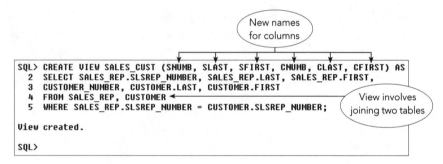

New names
for columns

```
SQL> CREATE VIEW SALES_CUST (SNUMB, SLAST, SFIRST, CNUMB, CLAST, CFIRST) AS
  2  SELECT SALES_REP.SLSREP_NUMBER, SALES_REP.LAST, SALES_REP.FIRST,
  3  CUSTOMER_NUMBER, CUSTOMER.LAST, CUSTOMER.FIRST
  4  FROM SALES_REP, CUSTOMER
  5  WHERE SALES_REP.SLSREP_NUMBER = CUSTOMER.SLSREP_NUMBER;

View created.

SQL>
```

View involves
joining two tables

Given the current data in the Premiere Products database, the SALES_CUST view contains the data shown in Figure 6.7.

FIGURE 6.7
···············
Using the
SALES_CUST
view

```
SQL> SELECT *
  2  FROM SALES_CUST;

SN SLAST        SFIRST    CNU CLAST        CFIRST
-- ----------   --------  --- ----------   --------
03 Jones        Mary      124 Adams        Sally
03 Jones        Mary      412 Adams        Sally
03 Jones        Mary      622 Martin       Dan
06 Smith        William   256 Samuels      Ann
06 Smith        William   315 Daniels      Tom
06 Smith        William   567 Dinh         Tran
06 Smith        William   587 Galvez       Mara
12 Diaz         Miguel    311 Charles      Don
12 Diaz         Miguel    405 Williams     Al
12 Diaz         Miguel    522 Nelson       Mary

10 rows selected.

SQL>
```

The next example involves statistics.

EXAMPLE 4 : Define a view named CRED_CUST that consists of each credit limit (CREDIT_LIMIT) and the number of customers who have this credit limit (NUMBER_CUSTOMERS).

The command shown in Figure 6.8 creates this view; the current data in the Premiere Products database also appears in the figure.

FIGURE 6.8
·················
Creating the
CRED_CUST
view

```
SQL> CREATE VIEW CRED_CUST (CREDIT_LIMIT, NUMBER_CUSTOMERS) AS
  2  SELECT CREDIT_LIMIT, COUNT(*)
  3  FROM CUSTOMER
  4  GROUP BY CREDIT_LIMIT;
                                          ◄──── View involves grouping

View created.

SQL>
SQL>
SQL>
SQL> SELECT *
  2  FROM CRED_CUST;

CREDIT_LIMIT NUMBER_CUSTOMERS
------------ ----------------
         750                2 ⎫
        1000                4 ⎬
        1500                3 ⎭──── Data in the view
        2000                1

SQL>
```

Using views has several benefits. First, views provide data independence. If the database structure changes (by adding columns, changing the way objects are related, etc.), the user still can access the same view. If adding extra columns to tables in the database is the only change and the user does not require these columns, you might not need to change the defining query for the view. If relationships change, the defining query might be different, but this difference is unknown to the user. The user continues to access the database through the same view, as though nothing has changed. For an example of the type of change that does require modification of the defining query for a view, suppose that customers are assigned to territories, that each territory is assigned to a single sales rep, that a sales rep can have more than one territory, and that a customer is represented by the sales rep who covers the customer's assigned territory. To implement these changes, you might choose to restructure the database as follows:

```
SALES_REP(SLSREP_NUMBER, LAST, FIRST, STREET, CITY, STATE, ZIP_CODE,
    TOTAL_COMMISSION, COMMISSION_RATE)
TERRITORY(TERRITORY_NUMBER, TERRITORY_DESCRIPTION, SLSREP_NUMBER)
CUSTOMER(CUSTOMER_NUMBER, LAST, FIRST, STREET, CITY, STATE, ZIP_CODE, BALANCE,
    CREDIT_LIMIT, TERRITORY_NUMBER)
```

Assuming that the SALES_CUST view shown earlier still is required, the defining query could be reformulated as follows:

```
CREATE VIEW SALES_CUST (SNUMB, SLAST, SFIRST, CNUMB, CLAST, CFIRST) AS
SELECT SALES_REP.SLSREP_NUMBER, SALES_REP.LAST, SALES_REP.FIRST,
CUSTOMER_NUMBER, CUSTOMER.LAST, CUSTOMER.FIRST
FROM SALES_REP, TERRITORY, CUSTOMER
WHERE SALES_REP.SLSREP_NUMBER = TERRITORY.SLSREP_NUMBER
AND TERRITORY.TERRITORY_NUMBER = CUSTOMER.TERRITORY_NUMBER;
```

The user of this view can still retrieve the number and name of a sales rep together with the number and name of every customer the sales rep represents. The user will be unaware, however, of the new structure in the database.

The second benefit of using views is that different users can view the same data in different ways because each user has his or her own view. In other words, the data can be customized to meet each user's needs.

The final benefit of using a view is that it can contain only those columns required by a given user. This practice accomplishes two things. First, because the view usually contains fewer columns than the overall database and because the view is a single table rather than a collection of tables, a view can greatly simplify the user's perception of the database. Second, views furnish a measure of security. Columns that are not included in the view are not accessible to the user. For example, omitting the BALANCE column from a view ensures that a user of the view cannot access any customer's balance. Likewise, rows that are not included in the view are not accessible. A user of the HOUSEWARES view, for example, cannot obtain any information about sporting goods, even though information about both housewares and sporting goods are stored in the same base table (PART).

These benefits hold true only when views are used for retrieval purposes. When updating the database, the issues involved in updating data through a view depend on the type of view, as you will see next.

Row-and-Column Subsets

Consider the row-and-column subset view for the HOUSEWARES view. There are columns in the underlying base table (PART) that are not present in the view. Thus, if you attempt to add a row with the data 'BB99','PAN',50,14.95, the system must determine how to enter the data in those columns from the PART table that are not included in the HOUSEWARES view (ITEM_CLASS and WAREHOUSE_NUMBER). In this case, it is clear what data to enter in the ITEM_CLASS column. According to the view definition, all rows are item class HW. But it is not clear what data to enter in the WAREHOUSE_NUMBER column. The only possibility would be NULL. Thus, provided that every column not included in a view can accept nulls, you can add new rows using the INSERT command. There is another problem, however. Suppose the user attempts to add a row containing the data 'AZ52','POT',25,9.95. This attempt *must* be rejected because there is a part number AZ52 that already exists in the PART table. Because this part is not in item class HW and not in the HOUSEWARES view, this rejection certainly will seem strange to the user, because there is no such part in the user's view.

On the other hand, updates or deletions cause no particular problem in this view. If the description of part number CA14 changes from Griddle to Pan, this change is made in the PART table. If part number CX11 is deleted, this deletion occurs in the PART table. One surprising change could take place, however. Suppose that the ITEM_CLASS column is included in the HOUSEWARES view and then a user changes the item class of part number CX11 from HW to AP. Because this item would no longer satisfy the criterion for being included in the HOUSEWARES view, part number CX11 would disappear from the user's view.

Whereas some problems do have to be overcome, it seems possible to update the database through the HOUSEWARES view. This does not imply that *any* row-and-column subset view is updatable, however. Consider the view shown in Figure 6.9. (The DISTINCT operator omits duplicate rows from the view.) This view currently contains the data shown in Figure 6.9.

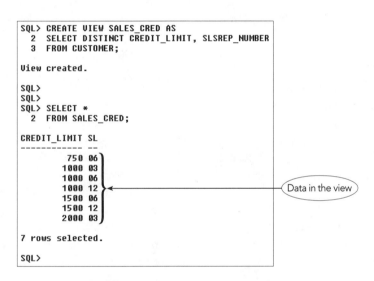

FIGURE 6.9

Creating the
SALES_CRED
view

```
SQL> CREATE VIEW SALES_CRED AS
  2  SELECT DISTINCT CREDIT_LIMIT, SLSREP_NUMBER
  3  FROM CUSTOMER;

View created.

SQL>
SQL>
SQL> SELECT *
  2  FROM SALES_CRED;

CREDIT_LIMIT SL
------------ --
         750 06
        1000 03
        1000 06
        1000 12       ← Data in the view
        1500 06
        1500 12
        2000 03

7 rows selected.

SQL>
```

How would you add the row 1000,'06' to this view? In the underlying base table (CUSTOMER), at least one customer must be added whose credit limit is $1,000 and whose sales rep number is 06, but who is it? You can't leave the other columns null in this case, because one of them is CUSTOMER_NUMBER, which is the base table's primary key. What would it mean to change the row 1200,'12' to 2000,'12'? Would it mean changing the credit limit to $2,000 for every customer represented by sales rep number 12 who currently has a credit limit of $1,200? Would it mean changing the credit limit of one of these customers and deleting the rest? What would it mean deleting the row 750,'06'? Would it mean deleting all customers whose credit limit is $750 and whose sales rep number is 06, or would it mean assigning these customers a different sales rep or a different credit limit? Potentially, you also could set the credit limit and/or the sales rep numbers to null.

Why does the SALES_CRED view involve a number of serious problems that are not present in the HOUSEWARES view? The basic reason is that the HOUSEWARES view includes, as one of its columns, the primary key of the underlying base table, but the SALES_CRED view does not. A row-and-column subset view that contains the primary key of the underlying base table is updatable (subject, of course, to some of the concerns already discussed).

Joins

In general, views that involve joins of base tables can cause problems at update. Consider the relatively simple SALES_CUST view, for example, described earlier (see Figures 6.6 and 6.7). The fact that some columns in the underlying base tables are not seen in this view presents some of the same problems discussed earlier. Even assuming that these problems can be overcome through the use of nulls, there are more serious problems inherent in the attempt to update the database through this view. On the surface, changing the row '06','Smith','William','256','Samuels','Ann' to '06','Baker','Nancy','256','Samuels','Ann' might not appear to pose any problems other than some inconsistency in the data. (In the new version of the row, the name of sales rep number 06 is Nancy Baker; in the next row in the table, the name of sales rep number 06, *the same sales rep*, is William Smith.)

The problem is actually more serious than that—making this change is not possible. The name of a sales rep is stored only once in the underlying SALES_REP table. Changing the name of sales rep number 06 from William Smith to Nancy Baker in this one row of the view causes the change to be made to the single row for sales rep number 06 in the SALES_REP table. Because the view simply displays data from the base tables, every row in which the sales rep number is 06 now shows the name as Nancy Baker. In other words, it appears that the same change has been made in all the other rows. In this case this change probably would be a good thing to ensure consistency in the data. In general, however, the unexpected changes caused by an update are not desirable.

Before concluding the topic of views that involve joins, you should note that all joins do not create the preceding problem. If two base tables happen to have the same primary key and the primary key used as the join column, updating the database is not a problem. For example, suppose that the actual database contains two tables (SLSREP_DEMO and SLSREP_FIN) instead of one table (SALES_REP). Figure 6.10 shows the data in these two tables.

FIGURE 6.10   SLSREP_DEMO and SLSREP_FIN tables

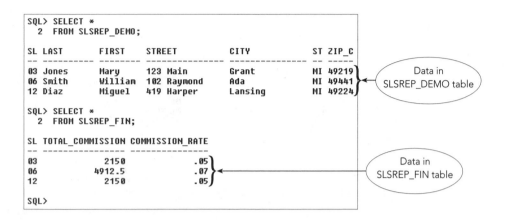

```
SQL> SELECT *
  2  FROM SLSREP_DEMO;

SL LAST       FIRST     STREET        CITY             ST ZIP_C
-- ---------- --------  ------------  ---------------- -- -----
03 Jones      Mary      123 Main      Grant            MI 49219
06 Smith      William   102 Raymond   Ada              MI 49441
12 Diaz       Miguel    419 Harper    Lansing          MI 49224

SQL> SELECT *
  2  FROM SLSREP_FIN;

SL TOTAL_COMMISSION COMMISSION_RATE
-- ---------------- ---------------
03             2150             .05
06           4912.5             .07
12             2150             .05

SQL>
```

Data in SLSREP_DEMO table

Data in SLSREP_FIN table

In this case, what was a single table in the Premiere Products design has been divided into two separate tables. Any user who expected to see a single table could be accommodated through a view that joins these two tables together using the SLSREP_NUMBER column. The view definition appears in Figure 6.11.

FIGURE 6.11   Creating the SLSREP view

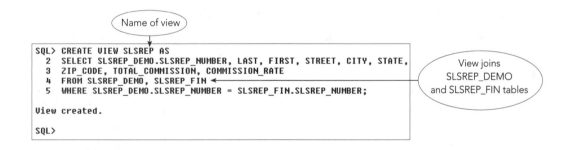

Name of view

```
SQL> CREATE VIEW SLSREP AS
  2  SELECT SLSREP_DEMO.SLSREP_NUMBER, LAST, FIRST, STREET, CITY, STATE,
  3  ZIP_CODE, TOTAL_COMMISSION, COMMISSION_RATE
  4  FROM SLSREP_DEMO, SLSREP_FIN
  5  WHERE SLSREP_DEMO.SLSREP_NUMBER = SLSREP_FIN.SLSREP_NUMBER;

View created.

SQL>
```

View joins SLSREP_DEMO and SLSREP_FIN tables

The SLSREP view appears in Figure 6.12.

FIGURE 6.12   Using the SLSREP view

Data in the view

```
SQL> SELECT *
  2  FROM SLSREP;

SL LAST        FIRST     STREET        CITY          ST ZIP_C TOTAL_COMMISSION COMMISSION_RATE
-- ----------  --------  ------------  ------------  -- ----- ---------------- ---------------
03 Jones       Mary      123 Main      Grant         MI 49219             2150             .05
06 Smith       William   102 Raymond   Ada           MI 49441            4912.5            .07
12 Diaz        Miguel    419 Harper    Lansing       MI 49224             2150             .05

SQL>
```

No difficulty is encountered in updating this view. To add a row, simply add a row to each underlying base table. To change data in a row, change the appropriate base table. To delete a row from the view, delete the corresponding rows from both underlying base tables.

QUESTION   How would you add the row '10','Peters', 'Jean','14 Brink', 'Holt', 'MI', '46223', 107.50,.05 to the SLSREP view?

ANSWER   Add the row '10','Peters', 'Jean', '14 Brink', 'Holt', 'MI', '46223' to the SLSREP_DEMO table and then add the row '10',107.50,.05 to the SLSREP_FIN table.

QUESTION   How would you change the name of sales rep number 03 to Mary Lewis?

ANSWER   Use an update to change the name in the SLSREP_DEMO table.

QUESTION   How would you change Mary's commission rate to .06?

ANSWER   Use an update to change the rate in the SLSREP_FIN table.

QUESTION   How would you delete sales rep number 06 from the SALES_REP table?

ANSWER   Delete sales rep number 06 from *both* the SLSREP_DEMO *and* SLSREP_FIN tables.

The SALES_REP view is updatable; updates (add, change, or delete) do not cause any problems. The main reason that this view is updatable—whereas other views involving joins are not—is that this view is derived from joining two base tables *on the primary key of each table*. In contrast, the SALES_CUST view is derived by joining two tables by matching the primary key of one table with a column that is *not* the primary key in the other table. Even more severe problems are encountered if neither of the join columns is a primary key column.

Statistics

A view that involves statistics calculated from one or more base tables is the most trouble-some view of all. Consider the CRED_CUST view, for example (see Figure 6.8). How would you add the row 900,3 to indicate that there are three customers who each have a credit limit of $900? Likewise, changing the row 750,2 to 750,5 means you are adding three new customers with credit limits of $750 each, for a total of five customers. Clearly these are impossible tasks; you can't add rows to a view that includes calculations. You only could add appropriate data to the underlying base tables.

Dropping a View

When a view is no longer needed, you can remove it by using the DROP VIEW command.

EXAMPLE 5 : The SLSREP view is no longer necessary, so remove it.

The DROP VIEW command to delete the SLSREP view is shown in Figure 6.13. The DROP VIEW command permanently deletes the SLSREP view from the database, but the tables and data on which the view is based still exist. The DROP VIEW command only removes the view definition.

FIGURE 6.13

Dropping a view

```
SQL> DROP VIEW SLSREP;

View dropped.

SQL>
```

◼ Security

Security is the prevention of unauthorized access to the database. Within an organization, the database administrator determines the types of access that various users need for the database. Some users might be able to retrieve and update data in the database. Other users might be able to retrieve any data from the database, but not make any changes to it. Still other users might be able to access only a portion of the database. For example, Bill can retrieve and update customer data, but cannot retrieve data for sales reps, orders, order lines, or parts. Mary can retrieve data only for parts. Sam can retrieve and update data on parts in item class HW, but cannot retrieve any data for other item classes.

Once determined, these access rules are enforced by whatever security mechanism the DBMS provides. SQL systems offer two security mechanisms. You already have seen that views furnish a certain amount of security. (If someone is accessing the database through a view, for example, he or she cannot access any data that is not part of the view.) The main mechanism, however, is the GRANT command. Here, the basic idea is that the database administrator can grant different types of privileges to users and then revoke them later, if necessary. These privileges include such things as the ability to select rows from a table, insert new rows, update existing rows, and so on. You can grant and revoke these privileges by using the GRANT and REVOKE commands. The following examples illustrate various uses of the GRANT command when the named users already exist in the database.

EXAMPLE 6 : User Jones must be able to retrieve data from the SALES_REP table.

The SQL statement to permit users to retrieve data includes the following SELECT statement for the table:

```
GRANT SELECT ON SALES_REP TO JONES;
```

EXAMPLE 7 : Users Smith and Brown must be able to add new parts to the PART table.

The SQL statement to add data includes the user names, separated by a comma, as follows:

```
GRANT INSERT ON PART TO SMITH, BROWN;
```

EXAMPLE 8 : User Anderson must be able to change the last name, first name, and street address of customers.

The SQL statement to update data includes the table name, followed by the column name(s) to update in parentheses, as follows:

```
GRANT UPDATE ON CUSTOMER (LAST, FIRST, STREET) TO ANDERSON;
```

EXAMPLE 9 : User Martin must be able to delete order lines.

The SQL statement to delete rows is as follows:

```
GRANT DELETE ON ORDER_LINE TO MARTIN;
```

EXAMPLE 10 : Every user must be able to retrieve part numbers, part descriptions, and item classes.

The SQL statement to indicate that all users have the privilege to retrieve data includes the special word PUBLIC, as follows:

```
GRANT SELECT ON PART (PART_NUMBER, PART_DESCRIPTION, ITEM_CLASS)
    TO PUBLIC;
```

EXAMPLE 11 : User Roberts must be able to create an index on the SALES_REP table.

You will learn about indexes and their uses in the next section. This example illustrates how to grant a user the ability to create an index. The SQL statement to create an index is as follows:

```
GRANT INDEX ON SALES_REP TO ROBERTS;
```

EXAMPLE 12 : User Thomas must be able to change the structure of the CUSTOMER table.

The SQL statement to change a table's structure is as follows:

```
GRANT ALTER ON CUSTOMER TO THOMAS;
```

EXAMPLE 13 : User Wilson must have all privileges for the SALES_REP, CUSTOMER, and ORDERS tables.

The SQL statement to indicate that a user has all privileges includes the use of the ALL privilege, as follows:

```
GRANT ALL ON SALES_REP, CUSTOMER, ORDERS TO WILSON;
```

The privileges that can be granted are SELECT to retrieve data, UPDATE to change data, DELETE to delete data, INSERT to add new data, INDEX to create an index, and ALTER to change the table structure.

The database administrator usually assigns privileges. Normally, when the database administrator grants a particular privilege to a user, the user cannot pass that privilege along to other users. If the user needs to be able to pass the privilege to another user, the GRANT statement must include the WITH GRANT OPTION clause. This clause grants the indicated privilege to the user and also permits the user to grant the same privileges (or a subset of them) to other users.

Any privileges granted can be revoked later by using the REVOKE command. The format of the REVOKE command is essentially the same as that of the GRANT command, but with two differences. Instead of GRANT <privilege> TO <user>, the format is REVOKE <privilege> FROM <user>. In addition, the clause WITH GRANT OPTION obviously is not meaningful as part of a REVOKE command. Incidentally, the revoke will cascade so that if Jones is granted privileges WITH GRANT OPTION and then Jones grants these same privileges to Smith, revoking the privileges from Jones revokes Smith's privileges at the same time. Example 14 illustrates the use of the REVOKE command.

EXAMPLE 14 : User Jones is no longer allowed to retrieve data from the SALES_REP table.

The SQL command to revoke a privilege is as follows:

```
REVOKE SELECT ON SALES_REP FROM JONES;
```

The GRANT and REVOKE commands also can be applied to views so that access is restricted to only certain rows within tables.

EXAMPLE 15 : Allow sales rep number 03 (Mary Jones) to access any data concerning the customers she represents, but do not permit her to access data concerning any other customer.

The SQL statement to restrict data access is as follows:

```
CREATE VIEW SLSR3CST AS
SELECT *
FROM CUSTOMER
WHERE SLSREP_NUMBER = '03'
GRANT SELECT ON SLSR3CST TO MARY JONES;
```

Indexes

Usually when you query a database you are searching for a row (or collection of rows) that satisfies some condition. Examining every row in a table to find the desired rows often takes too much time to be practical, especially if there are thousands of records in the table. Fortunately, you can create and use an **index** to speed up the searching process significantly. An index in a DBMS is similar to an index in a book. If you want to find a discussion of a given topic in a book, you can scan the entire book from start to finish and look for references to the topic. More than likely, however, you wouldn't resort to this time-consuming method. If the book has an index, you can use it to locate the page numbers on which your topic is discussed.

For relational model systems that run on both mainframes and microcomputers, the main mechanism for increasing the efficiency with which data is retrieved from the database is to use indexes. Consider Figure 6.14, for example, which shows the CUSTOMER table for Premiere Products together with one extra column named ROW_NUMBER. This extra column contains the row number of each row in the table (customer number 124 is in row 1, customer number 256 is in row 2, and so on). The DBMS uses these row numbers to go directly to a specific row. The users of the DBMS do not use the row numbers, and that is why usually you do not see them.

FIGURE 6.14   CUSTOMER table with row numbers

ROW_ NUMBER	CUSTOMER_ NUMBER	LAST	FIRST	STREET	CITY	STATE	ZIP_ CODE	BALANCE	CREDIT_ LIMIT	SLSREP_ NUMBER	CUSTOMER_ TYPE
1	124	Adams	Sally	481 Oak	Lansing	MI	49224	$ 818.75	$1,000	03	R
2	256	Jones	Ann	215 Pete	Grant	MI	49219	$ 21.50	$1,500	06	R
3	311	Charles	Don	48 College	Ira	MI	49034	$ 825.75	$1,000	12	R
4	315	Daniels	Tom	914 Cherry	Kent	MI	48391	$ 770.75	$ 750	06	R
5	405	Williams	Al	519 Watson	Grant	MI	49219	$ 402.75	$1,500	12	R
6	412	Adams	Sally	16 Elm	Lansing	MI	49224	$1817.50	$2,000	03	S
7	522	Nelson	Mary	108 Pine	Ada	MI	49441	$ 98.75	$1,500	12	R
8	567	Dinh	Tran	808 Ridge	Harper	MI	48421	$ 402.40	$ 750	06	R
9	587	Galvez	Mara	512 Pine	Ada	MI	49441	$ 114.60	$1,000	06	R
10	622	Martin	Dan	419 Chip	Grant	MI	49219	$1045.75	$1,000	03	D

To access a customer's row on the basis of his or her customer number, you might create and use an index, as shown in Figure 6.15. The index is a separate file that contains two columns. The first column contains a customer number, and the second column contains the number of the row in which the customer number is found. To find a customer, you look up the customer's number in the first column in the index. The value in the second column indicates which row to retrieve from the CUSTOMER table; then the row for the desired customer is retrieved.

FIGURE 6.15   Index for CUSTOMER table on CUSTOMER_NUMBER column

CUSTOMER_NUMBER INDEX

CUSTOMER_NUMBER	ROW_NUMBER
124	1
256	2
311	3
315	4
405	5
412	6
522	7
567	8
587	9
622	10

ROW_NUMBER	CUSTOMER_NUMBER	LAST	FIRST	STREET	CITY	STATE	ZIP_CODE	BALANCE	CREDIT_LIMIT	SLSREP_NUMBER	CUSTOMER_TYPE
1	124	Adams	Sally	481 Oak	Lansing	MI	49224	$ 818.75	$1,000	03	R
2	256	Jones	Ann	215 Pete	Grant	MI	49219	$ 21.50	$1,500	06	R
3	311	Charles	Don	48 College	Ira	MI	49034	$ 825.75	$1,000	12	R
4	315	Daniels	Tom	914 Cherry	Kent	MI	48391	$ 770.75	$ 750	06	R
5	405	Williams	Al	519 Watson	Grant	MI	49219	$ 402.75	$1,500	12	R
6	412	Adams	Sally	16 Elm	Lansing	MI	49224	$1817.50	$2,000	03	S
7	522	Nelson	Mary	108 Pine	Ada	MI	49441	$ 98.75	$1,500	12	R
8	567	Dinh	Tran	808 Ridge	Harper	MI	48421	$ 402.40	$ 750	06	R
9	587	Galvez	Mara	512 Pine	Ada	MI	49441	$ 114.60	$1,000	06	R
10	622	Martin	Dan	419 Chip	Grant	MI	49219	$1045.75	$1,000	03	D

Because customer numbers are unique, there will be a single row number in each row in the index. If the column on which the index is created is not unique, there might be multiple numbers in the rows in the index. Suppose, for example, that you need to access all customers who have a given credit limit. You also need to access all customers who are represented by a given sales rep. In this case, you might choose to create and use an index on the CREDIT_LIMIT column and an index on the SLSREP_NUMBER column, as shown in Figure 6.16. In the CREDIT_LIMIT index, the first column contains a credit limit, and the second column contains the numbers of *all* rows in which that credit limit is found. The SLSREP_NUMBER index is similar, except that the first column contains a sales rep number.

FIGURE 6.16   Index for CUSTOMER table on CREDIT_LIMIT column

CREDIT_LIMIT INDEX		SLSREP_NUMBER INDEX	
CREDIT_LIMIT	**ROW_NUMBERS**	**SLSREP_NUMBER**	**ROW_NUMBERS**
$ 750	4, 8	03	1, 6, 10
$1000	1, 3, 9, 10	06	2, 4, 8, 9
$1500	2, 5, 7	12	3, 5, 7
$2000	6		

ROW_ NUMBER	CUSTOMER_ NUMBER	LAST	FIRST	STREET	CITY	STATE	ZIP_ CODE	BALANCE	CREDIT_ LIMIT	SLSREP_ NUMBER	CUSTOMER_ TYPE
1	124	Adams	Sally	481 Oak	Lansing	MI	49224	$ 818.75	$1,000	03	R
2	256	Jones	Ann	215 Pete	Grant	MI	49219	$ 21.50	$1,500	06	R
3	311	Charles	Don	48 College	Ira	MI	49034	$ 825.75	$1,000	12	R
4	315	Daniels	Tom	914 Cherry	Kent	MI	48391	$ 770.75	$ 750	06	R
5	405	Williams	Al	519 Watson	Grant	MI	49219	$ 402.75	$1,500	12	R
6	412	Adams	Sally	16 Elm	Lansing	MI	49224	$1817.50	$2,000	03	S
7	522	Nelson	Mary	108 Pine	Ada	MI	49441	$ 98.75	$1,500	12	R
8	567	Dinh	Tran	808 Ridge	Harper	MI	48421	$ 402.40	$ 750	06	R
9	587	Galvez	Mara	512 Pine	Ada	MI	49441	$ 114.60	$1,000	06	R
10	622	Martin	Dan	419 Chip	Grant	MI	49219	$1045.75	$1,000	03	D

QUESTION   How would you use the index shown in Figure 6.16 to find every customer with a $1,500 credit limit?

ANSWER   Look up $1,500 in the CREDIT_LIMIT index to find a collection of row numbers (2, 5, and 7). Use these row numbers to find the corresponding rows in the CUSTOMER table (Ann Jones, Al Williams, and Mary Nelson).

QUESTION   How would you use the index shown in Figure 6.16 to find every customer who is represented by sales rep number 06?

ANSWER   Look up 06 in the SLSREP_NUMBER index to find a collection of row numbers (2, 4, 8, and 9). Use these row numbers to find the corresponding rows in the CUSTOMER table (Ann Jones, Tom Daniels, Tran Dinh, and Mara Galvez).

The actual structure of these indexes is more complicated than what is shown in the figures, but this representation is fine for our purposes. Fortunately, you don't have to be concerned with the details of manipulating and using indexes because the DBMS manages indexes for you. You decide which columns should have indexes built on them. Typically, you can create and maintain an index for any column or combination of columns in any table. Once an index has been created, the DBMS can use it to facilitate retrieval. No reference is made to any index by the user; rather, the DBMS automatically makes the decision whether to use a particular index.

As you would expect, using an index has advantages and disadvantages. An important advantage already mentioned is that an index makes certain types of retrieval more efficient. There are two disadvantages. First, an index occupies disk space that you could use for something else. Any retrieval that you can make using an index can also be made without the index; the index just speeds up the retrieval. The process of retrieval without using an index might be less efficient, but it still is possible. An index, in one sense, is technically unnecessary.

Second, the index must be updated whenever corresponding data in the database is updated. Without the index, these updates would not have to be performed. The main question to ask when considering whether to create a given index is: Do the benefits derived during retrieval outweigh the additional storage required and the extra processing involved in update operations? In a very large database, you might find that indexes are essential to decrease the time required to retrieve records. However, in a small database, an index might not provide any significant additional benefits.

You can add and drop indexes as needed. The final decision concerning the columns or combination of columns on which indexes should be built does not have to be made at the time the database is created. If the pattern of access to the database later indicates that overall performance would benefit from the creation of a new index, you can add an index. Likewise, if it appears that an existing index is unnecessary, you can drop it.

Creating an Index

Suppose that the users at Premiere Products find that they frequently need to display lists of customers that are ordered by the customers' balances. Also there are occasions when the users don't know a customer's number and need to locate the customer in the database by using the customer's name (last and first). Users also need to produce a report in which customers are listed by credit limit in descending order. Within the group of customers having the same credit limit, the customer records must be ordered by last name.

Each of the above requirements is carried out more efficiently when you create the appropriate index. The command used to create an index is CREATE INDEX, as illustrated in Example 16.

EXAMPLE 16 : Create an index named BALIND on the BALANCE column in the CUSTOMER table. Create an index named CUSTNAME on the combination of the LAST and FIRST columns in the CUSTOMER table. Create an index named CREDNAME on the combination of the CREDIT_LIMIT, LAST, and FIRST columns in the CUSTOMER table, with the credit limits listed in descending order.

The appropriate CREATE INDEX commands to create these indexes appear in Figure 6.17. Each command lists the name of the index and the table name on which the index is to be created. The column name(s) are listed in parentheses. If any column is to be included in descending order, the column name is followed by the word DESC.

FIGURE 6.17 Creating Indexes

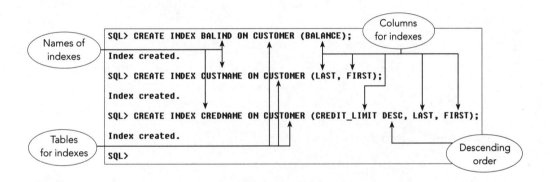

If customers are listed using the CREDNAME index, the records will appear in order of descending credit limit. Within any credit limit, the customers will be ordered by name.

Dropping an Index

The command used to drop (delete) an index is DROP INDEX. To delete the CREDNAME index, for example, the command would be as follows:

```
DROP INDEX CREDNAME;
```

Once the command is executed, the index no longer exists. CREDNAME was the index the DBMS used when listing customer records in descending credit limit order and then by customer name within credit limit. The DBMS still can list customers in this order; however, it cannot do so as efficiently as when an index could be used.

Unique Indexes

When you indicate the primary key for a table, the DBMS ensures automatically that the values for the primary key are unique. For example, the DBMS would automatically reject an attempt to add a second customer whose number is 124 because customer number 124 already exists. Thus, you don't need to take any special action to make sure that values in the primary key column are unique; the DBMS does it for you.

Occasionally, a column that is not the primary key might need unique values. For example, in the CUSTOMER table, the primary key is CUSTOMER_NUMBER. If the CUSTOMER table also contains a column for Social Security numbers, the values in this column also must be unique. Because the Social Security number column is not the table's primary key, however, you need to take special action in order for the DBMS to ensure that values in this column do not contain duplicates.

To ensure this uniqueness, create a special type of index called a **unique index** by using the CREATE UNIQUE INDEX command. To create a unique index named SSN on the SOC_SEC_NUMBER column of the CUSTOMER table, for example, the command would be as follows:

```
CREATE UNIQUE INDEX SSN ON CUSTOMER (SOC_SEC_NUMBER);
```

The unique index has all the properties of indexes already discussed, along with one additional property: the DBMS refuses to accept any update that would cause a duplicate value in the SOC_SEC_NUMBER column. In this case, that means the DBMS will reject the addition of a customer whose Social Security number is the same as that of another customer already in the database.

■ The System Catalog

Information concerning tables known to the DBMS is kept in the **system catalog**, or the **data dictionary**. In the following description of the catalog, the exact structure has been simplified, but it is representative of the basic idea.

The system catalog contains several tables of its own. You will focus on three important tables to simplify this discussion. The most common names for these three tables are SYSTABLES (information about the tables known to SQL), SYSCOLUMNS (information about the columns within these tables), and SYSVIEWS (information about indexes defined on these tables). Individual implementations of SQL might use different names for these tables. The examples that follow use Oracle to illustrate the concepts. In Oracle, the equivalent tables are named DBA_TABLES, DBA_TAB_COLUMNS, and DBA_VIEWS.

The system catalog is a relational database of its own. Consequently, in general, the same types of queries that are used to retrieve information from relational databases are used to retrieve information from the system catalog. The following examples illustrate this process.

Note: Usually users need special privileges to view the data in the system catalog. Thus, you might not be able to execute these commands. If you are executing the commands shown in the figures, substitute your user name for PRATT to list objects that you own. Your results will differ from those shown in the figures.

EXAMPLE 17 : List the name of every table for which the owner (creator of the table) is PRATT.

The command to list the table names owned by PRATT is shown in Figure 6.18. The WHERE clause restricts the table names to only those tables whose owner is PRATT.

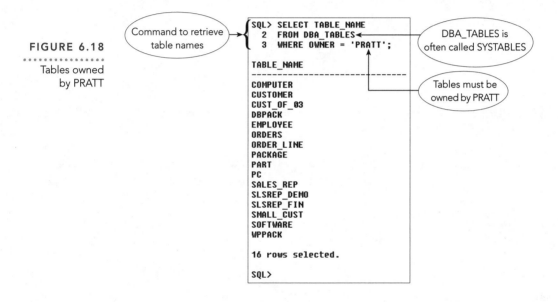

FIGURE 6.18
....
Tables owned
by PRATT

EXAMPLE 18 : List the name of every view whose owner is PRATT.

This command is similar to the command in Example 17. Rather than TABLE_NAME, the column to be selected is named VIEW_NAME. The command appears in Figure 6.19.

FIGURE 6.19
··················
Views owned
by PRATT

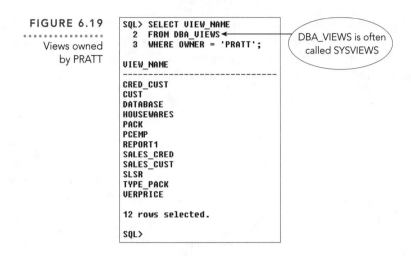

```
SQL> SELECT VIEW_NAME
  2   FROM DBA_VIEWS
  3   WHERE OWNER = 'PRATT';

VIEW_NAME
--------------------------------
CRED_CUST
CUST
DATABASE
HOUSEWARES
PACK
PCEMP
REPORT1
SALES_CRED
SALES_CUST
SLSR
TYPE_PACK
VERPRICE

12 rows selected.

SQL>
```

DBA_VIEWS is often called SYSVIEWS

EXAMPLE 19 : List every column and its associated data type in the CUSTOMER table whose owner is PRATT.

The command for this example appears in Figure 6.20. The columns to select are TABLE_NAME, COLUMN_NAME, and DATA_TYPE.

FIGURE 6.20   Columns in the CUSTOMER table

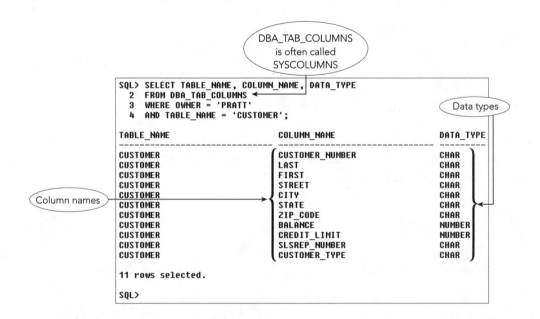

DBA_TAB_COLUMNS is often called SYSCOLUMNS

```
SQL> SELECT TABLE_NAME, COLUMN_NAME, DATA_TYPE
  2   FROM DBA_TAB_COLUMNS
  3   WHERE OWNER = 'PRATT'
  4   AND TABLE_NAME = 'CUSTOMER';

TABLE_NAME                      COLUMN_NAME                     DATA_TYPE
-----------------               ------------------------------  ---------
CUSTOMER                        CUSTOMER_NUMBER                 CHAR
CUSTOMER                        LAST                            CHAR
CUSTOMER                        FIRST                           CHAR
CUSTOMER                        STREET                          CHAR
CUSTOMER                        CITY                            CHAR
CUSTOMER                        STATE                           CHAR
CUSTOMER                        ZIP_CODE                        CHAR
CUSTOMER                        BALANCE                         NUMBER
CUSTOMER                        CREDIT_LIMIT                    NUMBER
CUSTOMER                        SLSREP_NUMBER                   CHAR
CUSTOMER                        CUSTOMER_TYPE                   CHAR

11 rows selected.

SQL>
```

Data types

Column names

140

EXAMPLE 20 : List every table owned by PRATT that contains a column named
CUSTOMER_NUMBER.

This command also uses the DBA_TAB_COLUMNS table as shown in Figure 6.21. In this case, the COLUMN_NAME column is used in the WHERE clause to restrict the rows to those in which the COLUMN_NAME is CUSTOMER_NUMBER.

FIGURE 6.21

Names of tables
owned by PRATT
that include the
CUSTOMER_
NUMBER column

```
SQL> SELECT TABLE_NAME
  2  FROM DBA_TAB_COLUMNS
  3  WHERE OWNER = 'PRATT'
  4  AND COLUMN_NAME = 'CUSTOMER_NUMBER';

TABLE_NAME
------------------------------
CUSTOMER
ORDERS
SMALL_CUST
CUST_OF_03

SQL>
```

You can obtain from the system catalog information about the tables in the relational database, the columns they contain, and the views built on them. You do this by using the same SQL syntax you use to query any other table.

Updating the tables in the system catalog occurs automatically when users create, alter, or drop tables, or when they create or drop indexes. Users should not update the catalog directly, using the update features of SQL, because inconsistent results might be produced. If a user deletes the row in the SYSCOLUMNS table for the CUSTOMER_NUMBER column, for example, the system would no longer have any knowledge of this column, which is the table's primary key; yet all the rows in the database would still contain a customer number. The system might now treat those customer numbers as last names, because as far as the system is concerned, the column named LAST is the first column in the CUSTOMER table.

■ Integrity in SQL

An **integrity constraint** is a rule that the data in the database must follow. Examples of integrity constraints in the Premiere Products database are as follows:

1. No two sales reps can have the same sales rep number.

2. The sales rep number for a customer must match the number of a sales rep currently in the database. For example, because there is no sales rep number 20, a customer cannot be assigned to sales rep number 20.

3. Item classes for parts must be AP, HW, or SG.

If a user enters data in the database that violates any of these constraints, the database will develop serious problems. For example, two sales reps with the same number, a customer with a nonexistent sales rep, or a part in a nonexistent item class would cause serious difficulties for users of the database. Thus, it is important to make sure that the data does not violate any integrity constraints. To solve these types of problems, the DBMS provides integrity support. **Integrity support** means that you can specify integrity constraints when creating your database, and the DBMS enforces these constraints. SQL has clauses to support three types of integrity constraints that you can specify within a CREATE TABLE command or an ALTER TABLE command. The only difference between these two commands is that an ALTER TABLE command is preceded by the word ADD to indicate that you are adding the constraint to any constraints that are already in place. To change an integrity constraint after it has been created, just enter the new constraint; the new constraint takes the place of the original.

The types of constraints supported in SQL are legal values, primary keys, and foreign keys. The CHECK clause ensures that only legal values that satisfy a particular condition are allowed in a given column. For example, to ensure that the only legal values for item class are AP, HW, or SG, use one of the following clauses:

```
CHECK (ITEM_CLASS IN ('AP', 'HW', 'SG') )
```

or

```
CHECK (ITEM_CLASS = 'AP' OR ITEM_CLASS = 'HW' OR ITEM_CLASS = 'SG')
```

The general form of the CHECK clause is simply the CHECK command followed by a condition. If any update to the database violates the condition, the update is rejected automatically.

The **primary key**, or the column or columns that uniquely identify a row in the table, is specified through the ADD PRIMARY KEY clause. For example, to indicate that SLSREP_NUMBER is the primary key for the SALES_REP table, the clause is as follows:

```
PRIMARY KEY (SLSREP_NUMBER)
```

In general, the PRIMARY KEY clause takes the form PRIMARY KEY followed by the column name in parentheses that makes up the primary key. If more than one column name is included, the column names are separated by commas.

A **foreign key** is a column in one table whose values match the primary key of another table. (One example is the SLSREP_NUMBER column in the CUSTOMER table. Values in this column are required to match those of the primary key of the SALES_REP table.) Any foreign keys are specified through FOREIGN KEY clauses. To specify a foreign key, you need to specify *both* the column that is a foreign key *and* the table it matches. In the CUSTOMER table, for example, SLSREP_NUMBER is a foreign key that must match the SLSREP_NUMBER value in the SALES_REP table. The clause to specify a foreign key is as follows:

```
FOREIGN KEY (SLSREP_NUMBER) REFERENCES SALES_REP
```

The general form for assigning a foreign key is: FOREIGN KEY, followed by the column name(s) that constitutes the foreign key, followed by the REFERENCES clause, and then by the table name that the foreign key is required to match.

EXAMPLE 21 : When the SALES_REP table was created, no primary key was assigned. Assign the SLSREP_NUMBER column as the primary key.

Figure 6.22 first describes the SALES_REP table and then shows the command to designate the primary key.

FIGURE 6.22 Assigning a primary key

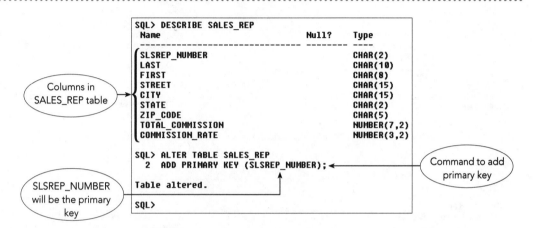

EXAMPLE 22 : Assign a foreign key for every table in the Premiere Products database.

Each foreign key to be specified requires a separate ADD FOREIGN KEY command. These commands appear in Figure 6.23.

FIGURE 6.23 Adding foreign keys

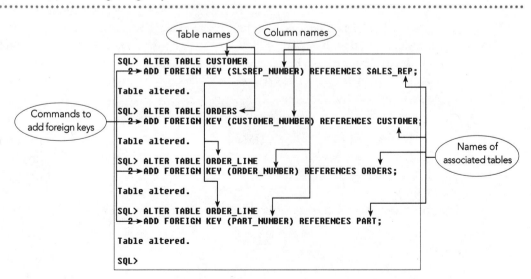

After creating the foreign keys, the DBMS will reject any update that violates the foreign key constraint. For example, the INSERT command shown in Figure 6.24 attempts to add an order for which the customer number (600) does not match any customer in the CUSTOMER table; thus the DBMS rejects the insert. The DELETE command in Figure 6.24 attempts to delete customer number 124. Deleting customer number 124 means that the rows in the ORDERS table for which the customer number is 124 no longer match any row in the CUSTOMER table; thus the DBMS rejects the deletion.

FIGURE 6.24   Violating foreign key constraints

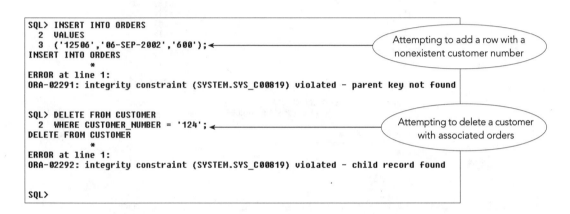

Note that the error messages shown in Figure 6.24 include the words "parent" and "child." When you use a foreign key, the table containing the foreign key is the **child** and the table referenced by the foreign key is the **parent**. For example, the CUSTOMER_NUMBER column in the ORDERS table is a foreign key that references the CUSTOMER table. For this foreign key, the CUSTOMER table is the parent and the ORDERS table is the child. The first error message indicates that there is no parent for the order (i.e., there is no customer number 600). The second error message indicates that there are child records (rows) for customer number 124 (i.e., customer number 124 has orders). The DBMS rejects both updates because they violate referential integrity.

EXAMPLE 23 : Specify that the units on hand values in the PART table must be greater than or equal to zero and less than or equal to 900. In addition, specify the valid item classes as AP, HW, or SG.

The commands to create these constraints appear in Figure 6.25.

FIGURE 6.25   Adding additional integrity constraints

```
SQL> ALTER TABLE PART
  2  ADD CHECK (UNITS_ON_HAND >= 0 AND UNITS_ON_HAND <= 900);

Table altered.

SQL> ALTER TABLE PART
  2  ADD CHECK (ITEM_CLASS IN ('AP', 'HW', 'SG') );

Table altered.

SQL>
```

Units on hand must be greater than or equal to zero and less than or equal to 900

Item class must be AP, HW, or SG

Now the DBMS will reject any update that violates either constraint. For example, it would reject the update shown in Figure 6.26 because the command attempts to change the item class to XX, which is an illegal value.

FIGURE 6.26   Violating an integrity constraint

```
SQL> UPDATE PART
  2  SET ITEM_CLASS = 'XX'
  3  WHERE PART_NUMBER = 'BT04';
UPDATE PART
*
ERROR at line 1:
ORA-02290: check constraint (SYSTEM.SYS_C00515) violated


SQL>
```

Attempting to set item class to XX

In this chapter, you learned about the purpose, creation, use, and benefits of views. Then you examined the features of SQL that relate to security, including granting and revoking various privileges that control the types of activities that users of the database need to perform. You learned about the purpose, advantages, and disadvantages of using indexes, and how to create and drop indexes. You saw how to obtain information from the system catalog. You also learned about the importance of integrity constraints, and how to create them. In the next chapter, you will learn how to use SQL commands to create reports.

■ SUMMARY

1. A view is a pseudotable whose contents are derived from data in existing base tables whenever users attempt to access the view.

2. To define a view, use the CREATE VIEW statement. This statement includes a defining query that describes the portion of the database included in the view. When a user retrieves data from the view, the query entered by the user is merged with the defining query, producing the query that SQL actually executes.

3. Views provide data independence, allow database access control, and simplify the database structure for users.

4. You cannot update views that involve statistics and views with joins of non-primary-key columns. In this case you must make all updates to the base table.

5. Use the DROP VIEW command to delete a view.

6. Use the GRANT command to give users access privileges to various portions of the database.

7. Use the REVOKE command to terminate previously granted privileges.

8. You can create and use an index to make data retrieval more efficient.

9. Use the CREATE INDEX command to create an index. Use the CREATE UNIQUE INDEX command to enforce a rule so that only unique values are allowed in a non-primary-key column.

10. Use the DROP INDEX command to delete an index.

11. The DBMS, not the user, makes the choice of which index to use to accomplish a given task.

12. The DBMS maintains information about the tables, columns, indexes, and other system elements in the system catalog. Information about tables is kept in the SYSTABLES table, information about columns is kept in the SYSCOLUMNS table, and information about views is kept in the SYSVIEWS table. In Oracle these same tables are named DBA_TABLES, DBA_TAB_COLUMNS, and DBA_VIEWS.

13. Use the SELECT command to obtain information from the system catalog. The DBMS updates the system catalog continuously; users do not update the catalog directly.

14. Integrity constraints are rules that the data in the database must follow to ensure that only legal values are accepted in specified columns, or that primary and foreign key values match between tables. To specify a general integrity constraint, use the CHECK clause. To specify a primary key, use the PRIMARY KEY clause. To specify a foreign key, use the FOREIGN KEY clause.

Use SQL to make the following changes to the Premiere Products database.

Note: If you are using Oracle for these exercises and wish to print a copy of your commands and results, type SPOOL followed by the name of a file, and then press the Enter key. All the commands from that point on will be saved in the file that you named. For example, to save the commands and results to a file named CHAPTER6.SQL on drive A, the command is as follows:

```
SPOOL A:CHAPTER6.SQL
```

When you have finished, type SPOOL OFF, and then press the Enter key to stop saving commands to the file. Then you can start any program that opens text files, open the file that you saved, and print it using the Print command on the File menu.

1. Define a view named SMALLCST. It consists of the customer number, last name, first name, street address, balance, and credit limit for every customer whose credit limit is $1,000 or less.
   a. Write the view definition for SMALLCST.
   b. Write a query to retrieve the customer number, last name, and first name of every customer in the SMALLCST view whose balance is over the credit limit.
   c. Write the query that the DBMS actually executes.
   d. Does updating the database through this view create any problems? If so, what are they? If not, why not?

2. Define a view named CUSTORD. It consists of the customer number, last name, first name, balance, order number, and order date for every order currently on file.
   a. Write the view definition for CUSTORD.

   b. Write a query to retrieve the customer number, last name, first name, order number, and order date for every order in the CUSTORD view for each customer whose balance is more than $500.
   c. Write the query that the DBMS actually executes.
   d. Does updating the database through this view create any problems? If so, what are they? If not, why not?

3. Define a view named ORDTOT. It consists of the order number and order total for each order currently on file. (The order total is the sum of the number of units ordered times the quoted price on each order line for each order.)
   a. Write the view definition for ORDTOT.
   b. Write a query to retrieve the order number and order total for every order whose total is over $500. Order the results by order number.
   c. Write the query that the DBMS actually executes.

d. Does updating the database through this view create any problems? If so, what are they? If not, why not?

4. Write the SQL commands to grant the following privileges:

a. User Stillwell must be able to retrieve data from the PART table.

b. Users Webb and Bradley must be able to add new orders and order lines.

c. User McKee must be able to change the number of units on hand for all parts.

d. User Thompson must be able to delete customers.

e. All users must be able to retrieve each customer's number, last name, first name, street address, city, state, and zip code.

f. User Pool must be able to create an index on the ORDERS table.

g. User Locke must be able to change the structure of the PART table.

h. User Scott must have all privileges on the ORDERS, ORDER_LINE, and PART tables.

i. User Richards must be permitted to access any data concerning housewares but not to access data concerning any other parts.

5. Write the SQL command to revoke user Stillwell's privilege.

6. Write the SQL commands to create the following indexes:

a. Create an index named PARTIND on the PART_NUMBER column in the ORDER_LINE table.

b. Create an index named PARTIND2 on the ITEM_CLASS column in the PART table.

c. Create an index named PARTIND3 on the ITEM_CLASS and WAREHOUSE_NUMBER columns in the PART table.

d. Create an index named PARTIND4 on the ITEM_CLASS and WAREHOUSE_NUMBER columns in the PART table, and list item classes in descending order.

7. Write the SQL command to delete the index named PARTIND3; it is no longer necessary.

8. Write the SQL commands to obtain the following information from the system catalog:

a. List every table that you own.

b. List every column in the PART table and its associated data type.

c. List every table that contains a column named PART_NUMBER.

d. List the name of every view in the system that is owned by you.

e. List the table name, column name, and data type for the columns named STREET, CITY, STATE, and ZIP_CODE. Order the results by table name.

9. Assume that the CUSTOMER table has been created, but that there are no integrity constraints. Write an ALTER TABLE command for the CUSTOMER table to ensure that the only values entered into the CREDIT_LIMIT column are 750, 1000, 1500, and 2000. The ALTER TABLE command should also indicate that the CUSTOMER_NUMBER column is the primary key and that the SLSREP_NUMBER column is a foreign key that must match the primary key of the table named SALES_REP.

▪EXERCISES (Henry Books)

Use SQL to make the following changes to the Henry Books database.

1. Define a view named BANTAM. It consists of the book code, book title, book type, and book price for every book published by Bantam Books.
   a. Write the view definition for BANTAM.
   b. Write a query to retrieve the book code, book title, and book price for every book with a price of less than $10.
   c. Write the query that the DMBS actually executes.
   d. Are there any problems created by updating the database through this view? If so, what are they? If not, why not?

2. Define a view named HARDBACK. It consists of the book code, book title, publisher name, and book price for every book that is not available in paperback.
   a. Write the view definition for HARDBACK.
   b. Write a query to retrieve the book title and publisher name for every book in the HARDBACK view that is priced at more than $20.
   c. Write the query that the DMBS actually executes.
   d. Are there any problems created by updating the database through this view? If so, what are they? If not, why not?

3. Define a view named VALUE. It consists of the branch number and a count of all books on hand for each branch.
   a. Write the view definition for VALUE.

   b. Write a query to retrieve the branch number and total count for the number of books in stock for each branch.
   c. Write the query that the DMBS actually executes.
   d. Are there any problems created by updating the database through this view? If so, what are they? If not, why not?

4. Write SQL commands to grant the following privileges:
   a. User Lopez must be able to retrieve data from the BOOK table.
   b. Users Bowen and Merrill must be able to add new books and publishers to the database.
   c. Users Jenkins and Sherman must be able to change the units on hand value.
   d. All users must be able to retrieve the book title, book code, and book price for every book.
   e. User Scout must be able to add and delete publishers.
   f. User Verner must be able to create an index for the BOOK table.
   g. Users Verner and Scout must be able to change the structure of the AUTHOR table.
   h. User Scout must have all privileges on all tables in the Henry Books database.
   i. User Chambers must be able to change the units on hand for books in branch number 2 but not to access data in any other branch.

5. User Verner has left the company. Write the SQL command to revoke all of her privileges.

6. Write SQL commands to create the following indexes:
   a. Create an index named BOOKIND on the BOOK_CODE column in the BOOK table.
   b. Create an index named BOOKIND2 on the BOOK_TYPE column in the BOOK table.
   c. Create an index named BOOKIND3 on the PUBLISHER_CODE and PUBLISHER_NAME columns in the PUBLISHER table, and list the publisher codes in descending order.

7. Write the SQL command to delete the index named BOOKIND3; it is no longer necessary.

8. Write the SQL commands to obtain the following information from the system catalog:
   a. List every table that you own that contains a column named BOOK_CODE or BRANCH_NUMBER. Order the results by table name.
   b. List every column in the PUBLISHER table and its associated data type.
   c. List every table that contains a column named PUBLISHER_CODE.
   d. List the name of every view.
   e. List the table name, column name, and data type for the columns named BOOK_CODE, BOOK_TITLE, and BOOK_PRICE. Order the results by column name.

9. Write the SQL commands to specify the following integrity constraints:
   a. Book code values must be less than 10,000.
   b. Book types must be PSY, FIC, HOR, MYS, ART, POE, SUS, MUS, or CS.
   c. The PAPERBACK column can accept only values of Y or N.
   d. The only branch numbers are 1, 2, 3, and 4.
   e. The only sequence numbers are 1 and 2.

10. Write the SQL commands to add the following foreign keys to the Henry Books database:
    a. PUBLISHER_CODE is a foreign key in the BOOK table.
    b. BRANCH_NUMBER is a foreign key in the INVENT table.
    c. AUTHOR_NUMBER is a foreign key in the WROTE table.

CHAPTER **7**

Reports

OBJECTIVES

- Use concatenation in a query
- Create a view for a report
- Create a query for a report
- Change column headings and formats in a report
- Add a title to a report
- Group data in a report
- Include totals and subtotals in a report
- Send a report to a file that can be printed

Introduction

In addition to the sorting and statistical functions available by using the SELECT command, many implementations of SQL include report formatting commands. In this chapter, you will learn about these commands by examining the report formatting commands found in Oracle.

In a sense, you can think of the output of a SELECT command as a very simple report that follows a rigid format. In many cases, this format is fine if you are concerned only about the particular data that appears in the result, and you are not concerned about the way the data is formatted.

If you need to generate a formatted report, however, you can't use a SELECT command. Fortunately, you can use other SQL commands to reformat the report in a variety of ways, and then you can rerun the query so the results are displayed in the desired format.

■ Concatenating Columns

Before creating a report, you need to understand the process of combining two or more character columns into a single expression because you will use this feature in the report that you will create later in this chapter. This process is called **concatenation**. To concatenate columns, you type two vertical lines (| |) between the column names, as illustrated in Example 1.

EXAMPLE 1   List the sales rep number and name of every sales rep. The name should be a concatenation of the FIRST and LAST columns.

The command and its results appear in Figure 7.1. The vertical lines between the column names in the SELECT command indicate a concatenation. The first and last names of each sales rep now appear in a single column in the query results.

FIGURE 7.1   Using concatenation

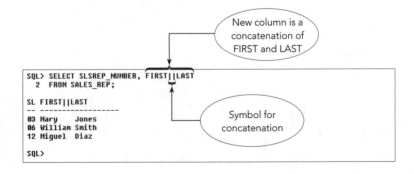

When the first name doesn't include sufficient characters to fill the width of the column (as is the case with Mary Jones and Miguel Diaz), SQL inserts extra spaces. To remove these extra spaces, you use the RTRIM (right trim) function. When you apply this function to the value in a column, SQL displays the original value and removes any spaces inserted at the end of the value. To use RTRIM in this query, the command is as follows:

```
SELECT SLSREP_NUMBER, RTRIM(FIRST)||' '||RTRIM(LAST)
FROM SALES_REP;
```

For sales rep number 03, for example, this command trims the first name to "Mary," concatenates it with a single space, and then concatenates the last name "Jones."

QUESTION   Why is it necessary to insert a single space character in single quotation marks in the query?

ANSWER   Without the space character, there would be no space between the first and last names. The name of sales rep number 03, for example, would be displayed as "MaryJones."

QUESTION   Is it necessary to trim the last name?

ANSWER   It is technically unnecessary to trim the end of the expression because you can't tell visually whether blanks are trimmed. However, trimming the last name in this case can be beneficial, especially when using the report formatting commands later in this chapter.

■ Creating and Using Scripts

When entering report formatting commands, it's a good idea to save the commands for future use. Otherwise, you must reenter the commands every time you want to produce the same report. Every version of SQL allows you to save commands. In this chapter, you will see how commands are saved in Oracle.

An Oracle file that contains SQL commands is called a **script**. You create a script by typing the EDIT command followed by the name of the script. For example, one of the scripts in this chapter is named FMT_RPT1. To create or edit this script, you would type EDIT FMT_RPT1 and then press the Enter key. (You can type the command in either uppercase or lowercase letters.)

The first time you type the EDIT command, the editor starts and asks if you want to create the file for the script. If you click the Yes button, the file is created and you can begin typing the necessary commands in the editor. If the file already exists, the file's contents are displayed, and then you can edit the file. In either case, when you close the editor, you are prompted to save your work. To run the commands in the script, type @ followed by the script's filename.

Note: You can save the script file to any storage location by including the drive and/or folder designation in the EDIT and @ commands. For example, to create the file named FMT_RPT1 in a folder named ORACLE on drive A, the command is EDIT A:\ORACLE\FMT_RPT1. The command to run the commands stored in this script is @ A:\ORACLE\FMT_RPT1.

Note: In some implementations of SQL, you must press the Enter key after typing the slash for the script to run correctly.

Note: If you receive an error message that your input is truncated when you try to run a script, edit your script to add a hard return after the slash.

Saving your commands in a script has another advantage—doing so allows you to develop your report in stages. You can create a file with an initial set of commands to format the report, and then see what the report looks like. If necessary, you can change the commands in the script to improve the report's appearance and then view the report again. You can make additional changes and view the report until you are satisfied with its appearance.

■ Creating a View for the Report

The data for a report can come from either a table or a view. It often is better to use a view, particularly if the report involves data from more than one table.

The report you will create in this chapter displays data from the SALES_REP and CUSTOMER tables. The report also concatenates two column names, as illustrated in Example 1. Therefore, it is advisable to create a view for the report. Example 2 creates this view.

EXAMPLE 2   Create a view for the report. The view should contain a column named SLSR that is the concatenation of the sales rep number, first name, and last name for every sales rep. There should be a hyphen between the sales rep number and name. Separate the first and last names by a single space. (For example, the resulting value in the SLSR column for the first sales rep should be "03-Mary Jones.") Format a second column named CUST in the same way, and display the customer number, first name, and last name for every customer. The view also should contain a column named BAL that contains the balance, a column named CRED that contains the credit limit, and a column named AVAIL that contains the available credit (CREDIT_LIMIT - BALANCE) for each customer.

The script containing the command to create this view appears in Figure 7.2. Notice that the command appears in Notepad, which is the default editor for Oracle. To open this window, type EDIT plus the name of the file (Cre_rpt1) at the SQL prompt, and then click the Yes button. The first line indicates that the view will consist of five columns named SLSR, CUST, BAL, CRED, and AVAIL. The third and fourth lines in the query specify a single column, SLSR, that uses concatenation and the RTRIM function. The fifth and sixth lines concatenate data into a single column named CUST. The remainder of the command is similar to many commands you have seen before.

FIGURE 7.2
Command to create the view for the report

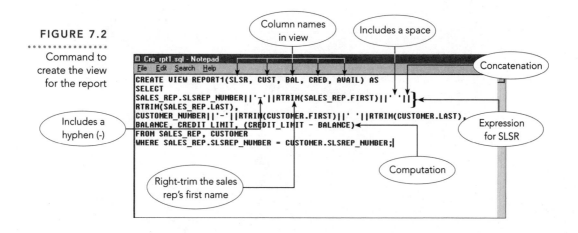

To save your changes to the script, close the editor and click the Yes button. The command shown in Figure 7.3 runs the script shown in Figure 7.2 and creates the view, which is named REPORT1. Once the CREATE VIEW command is executed, the view is ready to use.

FIGURE 7.3 Creating the view

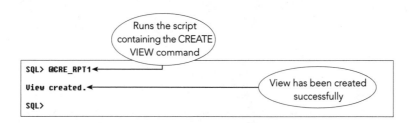

■ Executing a SELECT Command

To produce a report, you must run an appropriate SELECT command to create the data to use in the report. As a basis for this discussion on report formatting, you will use the query shown in Example 3.

⁞ List all the data in the REPORT1 view. Order the rows by the SLSR column.

The command to list the rows, which is stored in a script, appears in Figure 7.4.

FIGURE 7.4   Using the view

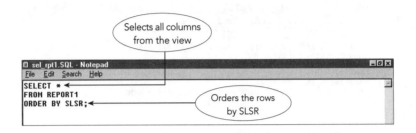

The query results appear in Figure 7.5. In the next examples, you will modify the format of the report to change the column headings, add a title, change the format of the numbers, and add totals and subtotals.

FIGURE 7.5   Executing the SELECT command

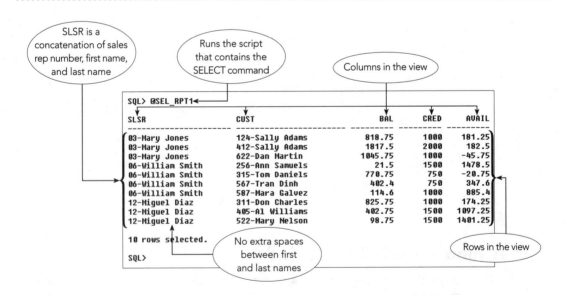

Note: The order of the output for the customer numbers within a given sales rep group might differ.

■ Changing Column Headings

The column headings shown in Figure 7.5 are not very descriptive of the columns' contents. You can change the headings to improve readability.

EXAMPLE 4 : Change the column headings in the report so they are more descriptive of the columns' contents.

To change a column heading, type the COLUMN command followed by the name of the column heading you want to change. Then use the HEADING clause to assign a new heading. If the heading is to extend over two lines, separate the two portions of the heading with a single vertical line (|). The commands to change the column headings appear in Figure 7.6.

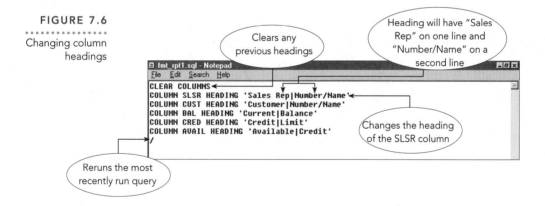

You can include the CLEAR COLUMNS command as a safety feature to clear any previous column changes. Without this command, any previous changes that you made to column headings or formats in the current work session would still be in place. The next commands make the required changes to the column headings. The slash (/) on the final line reruns the last query. This time, however, the results will reflect the changes to the column headings. The result of running the script containing these commands appears in Figure 7.7. Notice the new column headings.

FIGURE 7.7   Column headings changed

```
                Runs script to change                    New column
                 column headings                          headings

SQL> @FMT_RPT1

Sales Rep              Customer                   Current   Credit Available
Number/Name           Number/Name                Balance    Limit   Credit
--------------------- ----------------------- ---------- --------- ---------
03-Mary Jones         124-Sally Adams             818.75      1000    181.25
03-Mary Jones         412-Sally Adams             1817.5      2000     182.5
03-Mary Jones         622-Dan Martin             1045.75      1000    -45.75
06-William Smith      256-Ann Samuels               21.5      1500    1478.5
06-William Smith      315-Tom Daniels             770.75       750    -20.75
06-William Smith      567-Tran Dinh                402.4       750     347.6
06-William Smith      587-Mara Galvez              114.6      1000     885.4
12-Miguel Diaz        311-Don Charles             825.75      1000    174.25
12-Miguel Diaz        405-Al Williams             402.75      1500   1097.25
12-Miguel Diaz        522-Mary Nelson              98.75      1500   1401.25

10 rows selected.

SQL>
```

■ Changing Column Formats in a Report

You can use the COLUMN command to change more than just the column headings. You also can use the COLUMN command to change the width of a column or the way the entries appear in a column. Example 5 illustrates these types of formatting changes.

EXAMPLE 5   Change the format of the columns so the SLSR and CUST columns contain 18 characters each. The data in the other columns should be displayed with dollar signs and two decimal places.

The appropriate commands appear in Figure 7.8. The first two COLUMN commands change the format of the SLSR and CUST columns to A18. The letter A indicates that the column is alphanumeric (another name for character); the 18 indicates that the column is to be 18 characters long.

FIGURE 7.8   Changing formats

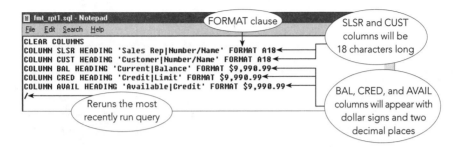

The next three COLUMN commands change the format of the three numeric columns in the view. In each case, the new format is $9,990.99. The 9s indicate that the value is numeric. The two 9s to the right of the decimal point indicate that each number will be displayed with two decimal places. The total number of 9s indicates the size of the column by representing the largest number that can be displayed. The dollar sign indicates that the values will be displayed as currency. Finally, the zero immediately to the left of the decimal point indicates that a value of zero will be displayed as $0.00. If you use all nines ($9,999.99), a value of zero will not be displayed; zero values will be left blank.

One way to construct the appropriate format for the numeric columns is to write the largest number that the column can display. If the amount can be over $10,000 but must be less than $100,000, for example, the format will be $99,999.99. If the amount must be less than $1,000 but more than $100, the format will be $999.99. The final step is to determine whether a value of zero should be displayed. If you want a zero value to be displayed, change the nine immediately to the left of the decimal to zero.

The result of running the script containing these commands appears in Figure 7.9. Notice the new format of the columns.

FIGURE 7.9   Report with formatting changes

```
SQL> @FMT_RPT1

Sales Rep            Customer               Current       Credit    Available
Number/Name          Number/Name            Balance        Limit       Credit
------------------   ------------------   ----------   ----------   ----------
03-Mary Jones        124-Sally Adams         $818.75    $1,000.00      $181.25
03-Mary Jones        412-Sally Adams       $1,817.50    $2,000.00      $182.50
03-Mary Jones        622-Dan Martin        $1,045.75    $1,000.00      -$45.75
06-William Smith     256-Ann Samuels          $21.50    $1,500.00    $1,478.50
06-William Smith     315-Tom Daniels         $770.75      $750.00      -$20.75
06-William Smith     567-Tran Dinh           $402.40      $750.00      $347.60
06-William Smith     587-Mara Galvez         $114.60    $1,000.00      $885.40
12-Miguel Diaz       311-Don Charles         $825.75    $1,000.00      $174.25
12-Miguel Diaz       405-Al Williams         $402.75    $1,500.00    $1,097.25
12-Miguel Diaz       522-Mary Nelson          $98.75    $1,500.00    $1,401.25

10 rows selected.

SQL>
```

Numbers are displayed as currency

■ Adding a Title to a Report

The next step is to add a title to the report, as illustrated in Example 6.

EXAMPLE 6 ⋮ Add a title to the report. The title should extend over two lines. The first line is "Customer Financial Report." The second line is "Organized by Sales Rep."

To add a title at the top of the report, use the TTITLE (top) command as shown in Figure 7.10. (To add a title at the bottom of the report, you use the BTITLE command.) Then you include the desired title in the TTITLE command. If the title will extend over two lines, separate the two portions by a vertical line, as shown in Figure 7.10.

FIGURE 7.10
· · · · · · · · · · · · · · · ·
Adding a title to
the top of a report

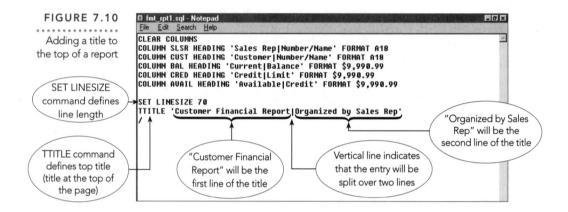

In order for the title to appear appropriately, you can adjust the line size by using the SET LINESIZE command. The **line size** determines where the title appears when it is centered across the line. In Figure 7.10, the SET LINESIZE command sets the line size to 70 characters. In this report, a line size of 70 characters is appropriate and places the title in the correct position on the line. In general, you can experiment with the line size to determine the best size for your report.

The resulting report appears in Figure 7.11. Notice that the page number and date appear automatically with the title.

FIGURE 7.11   Title added to the top of the report

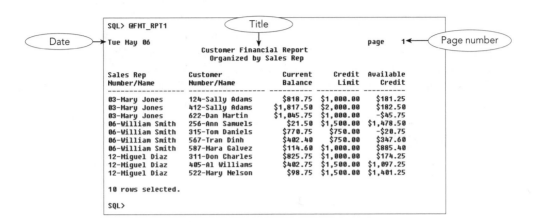

■ Grouping Data in a Report

Just as you can group data by using SQL queries, you also can group data in reports by using the BREAK command. You use the BREAK command to identify a column (or collection of columns) on which to group the data. The value in the column is displayed only at the beginning of the group. In addition, you can specify a number of lines to skip after each group. Example 7 illustrates the use of the BREAK command in grouping data. The example also removes the message "10 rows selected." from the end of the report.

EXAMPLE 7 : Group the rows in the report by the SLSR column. In addition, remove the message at the end of the report that indicates the number of rows selected.

To group rows by the SLSR column, the command is BREAK ON SLSR, as shown in Figure 7.12.

FIGURE 7.12

Adding a break

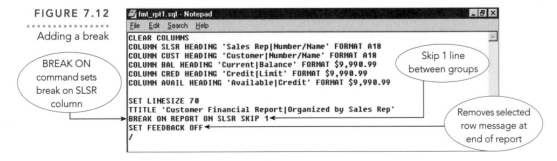

In order for the BREAK ON command to work properly, you need to sort the data on the indicated column. In this example, the data is sorted correctly because the SELECT command that produced the data included an ORDER BY SLSR clause.

The SKIP 1 clause at the end of the command indicates that one blank line should appear between groups. The SET FEEDBACK OFF command turns off the message indicating the number of rows selected by the query.

The results of executing these commands appear in Figure 7.13. Notice that the rows are grouped by sales rep, with the sales rep number and name appearing only once. Notice also that there is a blank line separating the groups. The message indicating the number of rows selected is not displayed.

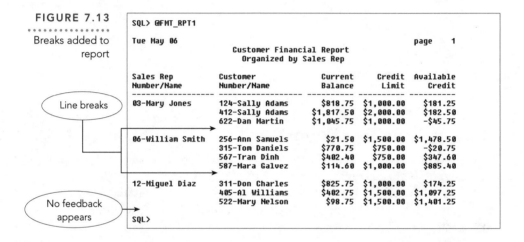

FIGURE 7.13

Breaks added to report

Line breaks

No feedback appears

```
SQL> @FMT_RPT1

Tue May 06                                                        page    1
                          Customer Financial Report
                          Organized by Sales Rep

Sales Rep           Customer             Current     Credit  Available
Number/Name         Number/Name          Balance      Limit     Credit
----------------    ----------------     -------    -------   --------
03-Mary Jones       124-Sally Adams      $818.75  $1,000.00    $181.25
                    412-Sally Adams    $1,817.50  $2,000.00    $182.50
                    622-Dan Martin     $1,045.75  $1,000.00    -$45.75

06-William Smith    256-Ann Samuels       $21.50  $1,500.00  $1,478.50
                    315-Tom Daniels      $770.75    $750.00    -$20.75
                    567-Tran Dinh        $402.40    $750.00    $347.60
                    587-Mara Galvez      $114.60  $1,000.00    $885.40

12-Miguel Diaz      311-Don Charles      $825.75  $1,000.00    $174.25
                    405-Al Williams      $402.75  $1,500.00  $1,097.25
                    522-Mary Nelson       $98.75  $1,500.00  $1,401.25

SQL>
```

■ Including Totals and Subtotals in a Report

A total that appears after each group is called a **subtotal**. In order to calculate a subtotal, you must include a BREAK command to group the rows. Then you can use a COMPUTE command to indicate the computation for the subtotal, as shown in Example 8.

EXAMPLE 8 : Include totals and subtotals for the BAL and AVAIL columns in the report.

The COMPUTE command uses **statistical functions** to calculate values to include in the report. The SQL statistical functions are shown in Table 7.1.

TABLE 7.1

Statistical
functions

Statistical function	Result of calculation
AVG	Average of values in a column
COUNT	Number of rows in a table
MAX	Largest value in a column
MIN	Smallest value in a column
STDEV	Standard deviation of values in a column
SUM	Sum of values in a column
VARIANCE	Variance of values in a column

In Example 8, SUM is the appropriate function to use. The script in Figure 7.14 shows the four COMPUTE commands needed to include totals and subtotals in the report.

FIGURE 7.14

Adding
computations
to a report

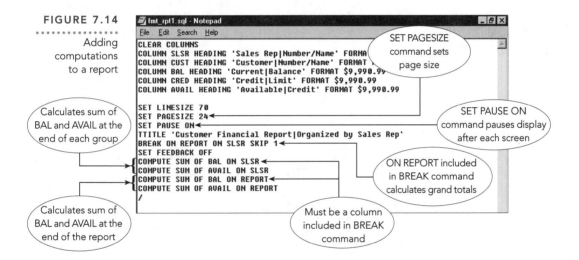

Notice that the commands in Figure 7.14 contain an OF clause that includes the desired computation and the column name on which the computation is to take place. The ON clause indicates the point at which the computation is to occur. The ON clause must contain the column name (or the word REPORT) that is included in the BREAK command.

The computations that end with ON REPORT represent computations that are displayed once at the end of the report. In this report, the computations represent grand totals. The computations associated with the other breaks occur at the end of the indicated group. In this report, the computations represent subtotals that are displayed after the group of customers of a particular sales rep.

There are two other commands in Figure 7.14 that are useful if the report will be displayed on the screen: SET PAGESIZE and SET PAUSE. In this example, the SET PAGESIZE command sets the number of lines that are displayed on a single page of the report to 24 (the screen line size). The SET PAUSE ON command causes the display of the report to pause after each screen. To display the next screen, press the Enter key. The result is shown in Figure 7.15.

FIGURE 7.15

Totals and subtotals included

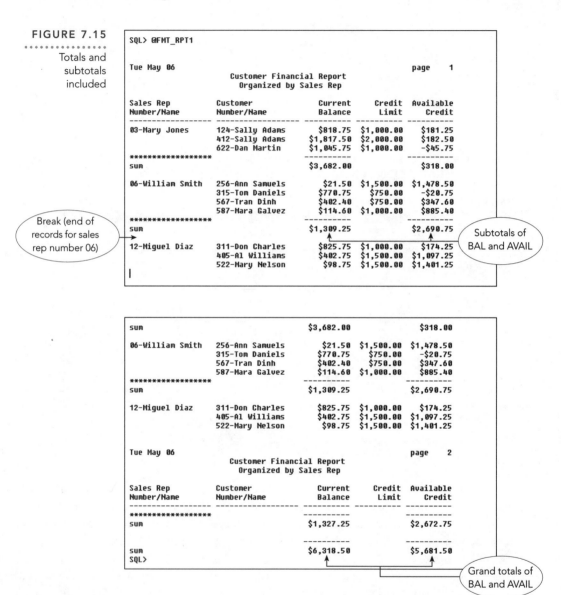

Break (end of records for sales rep number 06)

Subtotals of BAL and AVAIL

Grand totals of BAL and AVAIL

Note: In some SQL implementations, the SET PAUSE ON command causes the report to pause after you type the command to run the script. Press the Enter key to see the first and subsequent pages of the results.

The report includes subtotals to indicate the balances and available credit limits for every customer of each sales rep. Subtotals represent a subset of the overall total. Grand totals of the balance and available credit amounts for all customers appear at the end of the report.

Sending the Report to a File

In many cases, viewing the results of a query on the screen is sufficient. In other cases, you might want to print a copy; this is especially true for reports. The exact manner in which you print the report depends on the DBMS. If you have any questions about printing, consult your DBMS documentation.

To print a report using Oracle, you first send the output of the query to a file by using the SPOOL command. (The process of sending printed output to a file rather than directly to a printer is called **spooling**, and this is where the command gets its name.) After spooling your output to a file, you can print the contents of the file just as you would print the contents of any other file.

EXAMPLE 9   Send the report created in the previous examples to the file named REPORT1.SQL.

The SPOOL REPORT1.SQL command shown in Figure 7.16 begins sending the output of the subsequent commands to the file named REPORT1.SQL. The final command (SPOOL OFF) turns off spooling and stops any further output from being sent to the REPORT1.SQL file.

FIGURE 7.16   Sending the report to a file

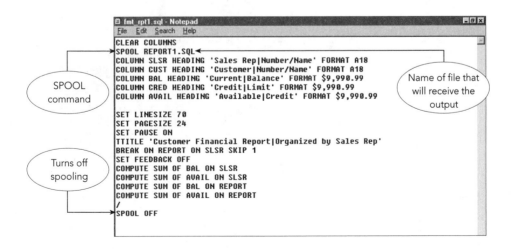

When you run the new script, the report is displayed again on the screen. As it is displayed on the screen (see Figure 7.17), it also is being written to the file. After completing the spooling process, the report is stored in the REPORT1.SQL file. Then you can print the file, edit it, include it in a document, or use it as needed in other ways.

FIGURE 7.17   Running the final report

```
Tue May 06                                              page    1
                      Customer Financial Report
                        Organized by Sales Rep

Sales Rep           Customer            Current   Credit  Available
Number/Name         Number/Name         Balance    Limit     Credit
------------------- ------------------- --------- -------- ----------
03-Mary Jones       124-Sally Adams      $818.75 $1,000.00   $181.25
                    412-Sally Adams    $1,817.50 $2,000.00   $182.50
                    622-Dan Martin     $1,045.75 $1,000.00   -$45.75
*******************                     --------           ----------
sum                                    $3,682.00            $318.00

06-William Smith    256-Ann Jones         $21.50 $1,500.00 $1,478.50
                    315-Tom Daniels      $770.75   $750.00   -$20.75
                    567-Tran Dinh        $402.40   $750.00   $347.60
                    587-Mara Galvez      $114.60 $1,000.00   $885.40
*******************                     --------           ----------
sum                                    $1,309.25          $2,690.75

12-Miguel Diaz      311-Don Charles      $825.75 $1,000.00   $174.25
                    405-Al Williams      $402.75 $1,500.00 $1,097.25
                    522-Mary Nelson       $98.75 $1,500.00 $1,401.25
*******************                     --------           ----------
sum                                    $1,327.25          $2,672.75

                                        ----------
sum                                    $6,318.50
SQL>
```

Report also sent to spool file

Note: You can save the file containing the report output by including the drive and/or folder names in the SPOOL command. For example, to save the file in a folder named ORACLE on drive A, the command is SPOOL A:\ORACLE\REPORT1.SQL.

■ A Complete Script to Produce the Report

You can include additional commands in the script to complete it. These commands are shown in Figure 7.18.

FIGURE 7.18   Completed script

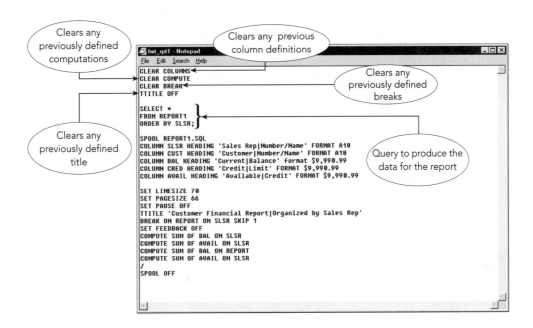

The first command, CLEAR COLUMNS, clears any previous column definitions. The next two commands have the same purpose. The CLEAR COMPUTE command clears any previously specified computations, and the CLEAR BREAK command clears any previous breaks. The TTITLE OFF command turns off any previously specified title.

The SQL query for the report is included next in the script. Without this query, a user who runs this script would have to execute the query first and then run the script. The script will not work if you don't execute the query, so it is a good idea to include the query execution command in the script to avoid this problem.

Because this report will be printed, it is appropriate to adjust the SET PAGESIZE and SET PAUSE commands. Change the page size to 66 (the length of a printed page) rather than 24 (the length of the screen). In addition, there is no reason for the display to pause for a report that will be printed, so change the SET PAUSE ON command to SET PAUSE OFF.

The remaining commands in the script are the same as those shown earlier. Running the script shown in Figure 7.18 produces the report shown in Figure 7.19.

FIGURE 7.19   Completed report

```
Tue May 06                                                   page    1
                          Customer Financial Report
                            Organized by Sales Rep

Sales Rep            Customer              Current    Credit   Available
Number/Name          Number/Name          Balance     Limit      Credit
------------------   ------------------   ---------- ---------- ----------
03-Mary Jones        124-Sally Adams        $818.75 $1,000.00    $181.25
                     412-Sally Adams      $1,817.50 $2,000.00    $182.50
                     622-Dan Martin       $1,045.75 $1,000.00    -$45.75
*****************                          ----------           ----------
sum                                       $3,682.00             $318.00

06-William Smith     256-Ann Samuels         $21.50 $1,500.00  $1,478.50
                     315-Tom Daniels        $770.75   $750.00    -$20.75
                     567-Tran Dinh          $402.40   $750.00    $347.60
                     587-Mara Galvez        $114.60 $1,000.00    $885.40
*****************                          ----------           ----------
sum                                       $1,309.25           $2,690.75

12-Miguel Diaz       311-Don Charles        $825.75 $1,000.00    $174.25
                     405-Al Williams        $402.75 $1,500.00  $1,097.25
                     522-Mary Nelson         $98.75 $1,500.00  $1,401.25
*****************                          ----------           ----------
sum                                       $1,327.25           $2,672.75

                                          ----------           ----------
sum                                       $6,318.50           $5,681.50
```

In this chapter, you learned about several important commands used to create and format reports. Table 7.2 shows these commands and their descriptions.

TABLE 7.2

· · · · · · · · · · · · · · · · ·

Reporting
command
summary

Command	Description
BREAK ON	Groups data in a report on a specified column
BTITLE	Adds a title at the bottom of a report
BTITLE OFF	Clears any previously specified title at the bottom of a report
CLEAR BREAK	Clears any previously specified report breaks
CLEAR COLUMNS	Clears any previous column changes
CLEAR COMPUTE	Clears any previously specified report computations
COLUMN	Changes the name of a column
COMPUTE	Calculates a count, minimum, maximum, sum, average, standard deviation, or variance on the values in a column in a report
HEADING	Assigns a new column heading
ON REPORT	Indicates that a calculation is to be performed on all values in the report
RTRIM	Deletes extra spaces that appear after a value in column
SET FEEDBACK OFF	Turns off the message indicating the number of rows selected by a query in a report
SET LINESIZE	Indicates the maximum number of characters on a line
SET PAGESIZE	Indicates the number of lines on a page
SET PAUSE	Indicates whether the screen display pauses after each screen of data
SKIP 1	Inserts one blank line between groups in a report
SPOOL	Sends query output to a file for printing or editing
TTITLE	Adds a title at the top of a report
TTITLE OFF	Clears any previously specified title at the top of a report

SUMMARY

1. To concatenate columns in report, separate the column names with two vertical lines (||). Use the RTRIM command to delete any extra spaces that follow the values.

2. Use a script to save the commands used to create a report so you can make modifications to the report at a later time. To save commands in a script, type EDIT followed by the script name. To run a script, type @ followed by the script name.

3. Reports are based on tables or views that contain the report data.

4. Use the COLUMN command to change a column heading. Use the HEADING clause to assign a new heading name. Type a single vertical line (|) to break a column heading over two lines.

5. Use the COLUMN command to change the format of column values.

6. Use the TTITLE or BTITLE command to add a title at the top or bottom of a report, respectively.

7. Use the BREAK command to group data in a report.

8. Use the BREAK and COMPUTE commands and an appropriate statistical function to calculate data in a report, such as totals and subtotals.

9. Use the SET PAUSE and SET PAGESIZE commands to pause a report after each screen of data, or to change the number of report lines to display.

10. Use the SPOOL command to send a report to a file for printing or editing.

■EXERCISES (Premiere Products)

Use SQL to make the following changes to the Premiere Products database.

Note: If you are using Oracle for these exercises and wish to print a copy of your commands and results, type SPOOL followed by the name of a file and then press the Enter key. All the commands from that point on are saved in the file that you named. For example, to save the commands and results to a file named CHAPTER7.SQL on drive A, the command is as follows:

```
SPOOL A:CHAPTER7.SQL
```

When you have finished, type SPOOL OFF, and then press the Enter key to stop saving commands to the file. Then you can start any program that opens text files, open the file that you saved, and print it using the Print command on the File menu.

1. Use SQL to produce a query that lists the name, street address, city, state, and zip code for every customer. Concatenate the first and last name data, and the city, state, and zip code data into single expressions. Insert a comma between the city and state data, and trim the columns so that only one space is displayed between them.

2. Create a view named REPORT2 for the query that you produced in Exercise 1.

Change the column headings to CUSTOMER_NAME, STREET_ADDRESS, and CITY_STATE_ZIP.

3. Change the column headings to CUSTOMER NAME, CUSTOMER ADDRESS, and CUSTOMER CITY/STATE/ZIP so that the word "CUSTOMER" appears on the first line and the other part of the heading appears on the second line.

4. Change the format of each column to 25 alphanumeric characters.

5. Create a view named REPORT3 that lists a trimmed concatenation of every customer's first and last name, along with his or her credit limit and current balance. The credit limit and balance should be displayed as currency. The column headings should be CUSTOMER NAME, CUSTOMER CREDIT LIMIT, and CUSTOMER BALANCE. The word "CUSTOMER" should be displayed on the first line, and the other part of the heading should be displayed on the second line.

6. Add the title "CUSTOMER CREDIT LIMITS AND BALANCES" to the report. The title should be displayed on two lines, with the words "AND BALANCES" on the second line.

7. Change the report so that the feedback about the number of rows selected is not displayed.

8. Write a script that creates the following report: List the number and name of every sales rep together with the number, name, and balance of every customer represented by the sales rep. Column headings for these columns are SNUM, NAME, CNUM, NAME, and BAL, respectively. Sort the report by customer number within sales rep number. Format the BALANCE column with two decimal places, and total it by sales rep. A sales rep number and name should appear only once. The heading for the report is "SALES REP REPORT." Format the report so that it can be printed with 66 lines on the page and no screen pauses. (*Hint*: Create a view to select the data.)

■EXERCISES (Henry Books)

Use SQL to make the following changes to the Henry Books database.

1. Create the following report: For every book published, list the publisher name and a concatenation of the publisher's city and state. Then list the book title, selling price, and number of units on hand by branch number. Column headings for these columns are PUBLISHER NAME, PUBLISHER LOCATION, BOOK TITLE, PRICE, UNITS ON HAND, and BRANCH NUMBER. The title of the report is "INVENTORY LIST FOR HENRY BOOKS," printed on two lines. Format the columns that contain currency as such. Group the report by branch number. Format the report so that it can be printed with 66 lines on the page and with no screen pauses.

Embedded SQL

OBJECTIVES

- Embed SQL commands in COBOL statements

- Retrieve single rows using embedded SQL

- Update a table using embedded INSERT, UPDATE, and DELETE commands

- Use cursors to retrieve multiple rows in embedded SQL

- Update a database using cursors

- Learn how to handle errors in programs containing embedded SQL commands

Note: Chapter 8 is optional—you must be familiar with COBOL or another procedural programming language in order to understand the contents of this chapter.

Introduction

SQL is a very powerful non-procedural language in which you communicate tasks to the computer using simple commands. As in other non-procedural languages, you can accomplish many tasks by using a single, relatively simple command. By contrast, a procedural language is one in which you must give the computer the step-by-step process for accomplishing tasks. Accomplishing a task in a procedural language might require many lines of code. COBOL is an example of a procedural language. This chapter uses COBOL to illustrate how to embed SQL commands into another language.

Even though SQL and other non-procedural languages are well-equipped to store and query data, sometimes you might need to complete tasks that are beyond the capabilities of SQL. In such cases, you need to use a procedural language. Fortunately, you can embed SQL commands into a procedural language to capitalize on the advantages of SQL. You can use SQL for some tasks, and then include embedded SQL commands in the procedural language to accomplish tasks that are beyond the capabilities of SQL.

In this chapter, you will learn how to embed SQL commands in COBOL. The process of embedding SQL in other programming languages, such as C or PL/SQL, is very similar. Because the focus of this chapter is not on the COBOL language but rather is on the use of SQL within COBOL, it uses simple ACCEPT and DISPLAY statements within the COBOL programs for input and output.

A COBOL program in which SQL commands are embedded will have additional statements beyond the standard COBOL statements in both the DATA and PROCEDURE divisions. In both cases, these new statements are preceded by EXEC SQL and followed by END-EXEC, so that the COBOL compiler can distinguish embedded SQL commands from standard COBOL statements.

Note: Your instructor will inform you of other requirements for the way you embed SQL commands.

In the DATA DIVISION (in particular, in the WORKING-STORAGE SECTION), the new statements declare the tables that will be used in processing the database as well as define a communications area for SQL. The **communications area** includes items that allow SQL to communicate various aspects of processing to the program. The main item you use in the communications area is SQLCODE. After executing any SQL statement, SQLCODE will contain a code indicating the fate of the executed statement. If the execution is normal, SQLCODE is zero. If the execution is not normal, the value in SQLCODE indicates the problem that occurred (for example, not finding any rows that satisfy the condition in a WHERE clause). Programs should contain statements that check the value of SQLCODE after each SQL statement is executed.

In the PROCEDURE DIVISION, the new statements will be SQL statements, with some slight variations. The examples that follow illustrate how to use SQL to retrieve a single row, insert new rows, and update and delete existing rows. Finally, you will learn how to retrieve multiple rows. Executing a SELECT statement that retrieves more than one row presents a problem for a language like COBOL, which is oriented toward processing one record at a time. Thus, you must take some special action in such situations.

■ DATA DIVISION

Any tables to be processed in COBOL must be declared in WORKING-STORAGE, the portion of the program where you declare your variables. To do this, you use the DECLARE TABLE command, which is similar to the SQL CREATE TABLE command. To process the SALES_REP table, for example, the code is written as follows:

```
EXEC SQL
    DECLARE SALES_REP TABLE
        (SLSREP_NUMBER DECIMAL (2),
        LAST                CHAR (10),
        FIRST               CHAR (8),
        STREET              CHAR (15),
        CITY                CHAR (15),
        STATE               CHAR (2),
        ZIP_CODE            CHAR (5),
        TOTAL_COMMISSION    DECIMAL (7,2),
        COMMISSION_RATE     DECIMAL (3,2) )
END-EXEC.
```

Optionally, if the description of the SALES_REP table is stored in a special location, often called a **library**, under the name DECSALES_REP, the code is written as follows:

```
EXEC SQL
    INCLUDE DECSALES_REP
END-EXEC.
```

When processing this table, you will need to use regular COBOL variables corresponding to the columns in the table. For the SALES_REP table, for example, the COBOL code is written as follows:

```
01 W-SALES-REP.
        03 W-SLSREP-NUMBER      PIC S9(2)        COMP-3.
        03 W-LAST               PIC X(10).
        03 W-FIRST              PIC X(8).
        03 W-STREET             PIC X(15).
        03 W-CITY               PIC X(15).
        03 W-STATE              PIC X(2).
        03 W-ZIP-CODE           PIC X(5).
        03 W-TOTAL-COMMISSION   PIC S9(5)V9(2)   COMP-3.
        03 W-COMMISSION-RATE    PIC S9V9(2)      COMP-3.
```

Because this description is standard COBOL, you do not precede the code with EXEC SQL.

Note: The programs in this chapter use a naming approach in which the work variables begin with the letter "W" and are followed by a hyphen. This naming convention emphasizes that the variables are work variables and not items from the database.

Finally, the system uses the **SQL communication area** (**SQLCA**) to provide feedback to the program using SQLCODE. You include the SQLCA by coding the command as follows:

```
EXEC SQL
     INCLUDE SQLCA
END-EXEC.
```

There is only one other new type of entry—a cursor—that appears in the DATA DIVISION. You use a **cursor** for the multiple-row SELECT statements mentioned earlier. Cursors and the problems associated with multiple-row retrieval are discussed later in this chapter.

■ PROCEDURE DIVISION

Before looking at examples of SQL statements in the PROCEDURE DIVISION, some general comments are necessary. First, you can use normal COBOL variables in SQL statements. Such variables are called **host variables**; they are variables in the host language, in this case COBOL. When used, you must precede the variable with a colon. If you use W-LAST *within a SQL statement*, for example, you must enter it as :W-LAST. For any other use, it can appear as W-LAST. Second, you must place the results of SQL queries in host variables in the INTO clause, which is written as follows:

```
SELECT LAST
     INTO :W-LAST
     FROM SALES_REP
     WHERE SLSREP_NUMBER = "03"
```

Note: When naming host variables, you must follow the naming rules in the host language. In COBOL, for example, hyphens are allowed, but underscores are not. One way to avoid problems is to use hyphens in the host variable wherever there is an underscore in the corresponding SQL column name. For example, the host variable corresponding to the SQL ZIP_CODE column becomes W-ZIP-CODE. Notice that the underscore separating the words ZIP and CODE has changed to a hyphen in COBOL, and that the COBOL variable is preceded by "W-".

Finally, you must make provisions for exceptional conditions. Such conditions occur when no data is found to satisfy a condition or when no space is available to add a new row on the disk. The specific condition determines which action occurs. Mechanisms to check for and handle these conditions are discussed after the following examples.

Retrieve a Single Row and Column

Example 1 illustrates using embedded SQL to retrieve a single row and column from a table.

EXAMPLE 1 : Obtain the last name of sales rep number 03 and place it in W-LAST.

Because this retrieval is based on the primary key (SLSREP_NUMBER), it does not pose any problem for a record-at-a-time language like COBOL. If SQL is used in a stand-alone mode, the statement is written as follows:

```
SELECT LAST
    FROM SALES_REP
    WHERE SLSREP_NUMBER = "03";
```

In COBOL, the statement is written as follows:

```
EXEC SQL
    SELECT LAST
        INTO :W-LAST
        FROM SALES_REP
        WHERE SLSREP_NUMBER = "03"
END-EXEC.
```

The only difference between the statements—other than the required EXEC SQL and END-EXEC codes—is the addition of the INTO clause indicating that the result should be placed in the host. The W-LAST variable now can be used in any way it could be used in any other COBOL program. Its value could be printed in a report, displayed on screen, compared with some other name, and so on.

Retrieve a Single Row and All Columns

There are two differences between Example 2 and Example 1. First, Example 2 requires retrieving all columns. Second, the sales rep number is not given in this example; it is stored in a host (COBOL) variable.

EXAMPLE 2 : Obtain all information about the sales rep whose number is stored in the host variable W-SLSREP-NUMBER.

To fill in W-SLSREP-NUMBER, you can use an appropriate COBOL statement, such as MOVE. This statement copies the value in one variable into another. You also can use ACCEPT, which obtains input from a user and places it in a variable. Then you can use the following embedded SQL command:

```
EXEC SQL
    SELECT LAST, FIRST, STREET, CITY, STATE, ZIP_CODE,
        TOTAL_COMMISSION, COMMISSION_RATE
        INTO :W-LAST, :W-FIRST, :W-STREET, :W-CITY, :W-STATE,
            :W-ZIP-CODE, :W-TOTAL-COMMISSION,
            :W-COMMISSION-RATE
        FROM SALES_REP
        WHERE SLSREP_NUMBER = :W-SLSREP-NUMBER
END-EXEC.
```

In this formulation, several columns are listed in the SELECT statement, and the corresponding host variables that will receive the values are listed in the INTO statement. In addition, the host variable W-SLSREP-NUMBER is used in the WHERE clause. Note that there was no need to select SLSREP_NUMBER and place it in W-SLSREP-NUMBER, because W-SLSREP-NUMBER already contains the desired number.

Figure 8.1 shows a complete COBOL program to accomplish the task. The numbered portions of the program are described in the section following the figure.

FIGURE 8.1   Program to display sales rep information

```
                      IDENTIFICATION DIVISION.
                      PROGRAM-ID.     DSPSLS.

                      ENVIRONMENT DIVISION.

                      DATA DIVISION.

                      WORKING-STORAGE SECTION.

     1    01  STATUS-FLAGS.
               03  ARE-WE-DONE              PIC X(3).
                    88  WE-ARE-DONE                     VALUE 'YES'.

     2    01  W-SALES-REP.
               03  W-SLSREP-NUMBER          PIC X(2).
               03  W-LAST                   PIC X(10).
               03  W-FIRST                  PIC X(8).
               03  W-STREET                 PIC X(15).
               03  W-CITY                   PIC X(15).
               03  W-STATE                  PIC X(2).
               03  W-ZIP-CODE               PIC X(5).
               03  W-TOTAL-COMMISSION       PIC S9(5)V9(2).
               03  W-COMMISSION-RATE        PIC SV9(2).

     3         EXEC SQL
                 DECLARE SALES_REP TABLE
                       (SLSREP_NUMBER      CHAR (2),
                        LAST               CHAR (10),
                        FIRST              CHAR (8),
                        STREET             CHAR (15),
                        CITY               CHAR (15),
                        STATE              CHAR (2),
                        ZIP_CODE           CHAR (5),
                        TOTAL_COMMISSION   DECIMAL (7,2),
                        COMMISSION_RATE    DECIMAL (3,2) )
                 END-EXEC.

     4         EXEC SQL
                    INCLUDE SQLCA
               END-EXEC.

               PROCEDURE-DIVISION.

               MAIN-PROGRAM.
                  MOVE 'NO' TO ARE-WE-DONE.
                  PERFORM MAIN-LOOP
                      UNTIL WE-ARE-DONE.
                  STOP RUN.

               MAIN-LOOP.
     5            DISPLAY 'SALES REP NUMBER OR ** TO END PROGRAM: '.
                  ACCEPT W-SLSREP-NUMBER.
                  IF W-SLSREP-NUMBER = '**'
                      MOVE 'YES' TO ARE-WE-DONE
                   ELSE
                      PERFORM FIND-AND-DISPLAY-SALES_REP.

               FIND-AND-DISPLAY-SALES_REP.
     6         EXEC SQL
                    SELECT LAST, FIRST, STREET, CITY, STATE, ZIP_CODE,
                         TOTAL_COMMISSION, COMMISSION_RATE
                    INTO :W-LAST, :W-FIRST, :W-STREET, :W-CITY, :W-STATE,
                         :W-ZIP_CODE, :W-TOTAL-COMMISSION,
                         :W-COMMISSION-RATE
                    FROM SALES_REP
                    WHERE SLSREP_NUMBER = :W-SLSREP-NUMBER
               END-EXEC.
     7         IF SQLCODE = 0
                    DISPLAY '      LAST NAME: ', W-LAST,
                    DISPLAY '     FIRST NAME: ', W-FIRST,
                    DISPLAY '         STREET: ', W-STREET,
                    DISPLAY '           CITY: ', W-CITY,
                    DISPLAY '          STATE: ', W-STATE,
                    DISPLAY '       ZIP CODE: ', W-ZIP-CODE
                    DISPLAY 'TOTAL COMMISSION: ', W-TOTAL-COMMISSION
                    DISPLAY ' COMMISSION RATE: ', W-COMMISSION-RATE
                 ELSE
                    DISPLAY 'THERE IS NO SUCH SALES REP'.
```

1. This program contains a loop. The user can display sales reps repeatedly. The data item named ARE-WE-DONE is a flag that indicates that the user does not wish to display any more sales reps. (The user indicates that he or she is finished by entering a sales rep number of zero.)

2. The record named W-SALES-REP contains the host variables that store the sales rep data. Within W-SALES-REP, there is a field for each column in the SALES_REP table. The approach used here for naming these fields is to precede the name of the column in the table with "W-".

3. This statement is the declaration of the SALES_REP table. Note that it is very similar to the SQL CREATE TABLE command.

4. This statement is used to include the SQL communication area in the program.

5. This is the main loop where the user is asked to enter the number of the desired sales rep or zero if no more sales reps are desired. If the user enters zero, the flag ARE-WE-DONE is set to YES, indicating that the process should terminate. If not, the program performs the paragraph named FIND-AND-DISPLAY-SALES_REP.

6. This embedded SELECT statement selects the desired sales rep and places the information about the sales rep in the indicated host variables.

7. If SQLCODE contains the number zero, the desired sales rep was found successfully, in which case the information is displayed. If not, the sales rep is not in the database, and an appropriate error message is displayed.

Figure 8.2 represents a slightly different version of the same program. The only difference here occurs on line 1. Instead of including the specific table declaration for SALES_REP in the program, as in the previous version, the INCLUDE DECSALES_REP statement is included. This statement assumes that the declaration has been created already and is stored in the file DECSALES_REP.

FIGURE 8.2   Program to display sales rep information (version 2)

```
                        IDENTIFICATION DIVISION.
                        PROGRAM-ID.    DSPSLS.

                        ENVIRONMENT DIVISION.

                        DATA DIVISION.

                        WORKING-STORAGE SECTION.

                        01  STATUS-FLAGS.
                            03  ARE-WE-DONE            PIC X(3).
                                88  WE-ARE-DONE                      VALUE 'YES'.

                        01  W-SALES-REP.
                            03  W-SLSREP-NUMBER        PIC X(2).
                            03  W-LAST                 PIC X(10).
                            03  W-FIRST                PIC X(8).
                            03  W-STREET               PIC X(15).
                            03  W-CITY                 PIC X(15).
                            03  W-STATE                PIC X(2).
                            03  W-ZIP-CODE             PIC X(5).
                            03  W-TOTAL-COMMISSION     PIC S9(5)V9(2).
                            03  W-COMMISSION-RATE      PIC SV9(2).

         1              EXEC SQL
                            INCLUDE DECSALES_REP
                        END-EXEC.

                        EXEC SQL
                            INCLUDE SQLCA
                        END-EXEC.

                        PROCEDURE-DIVISION.

                        MAIN-PROGRAM.
                            MOVE 'NO' TO ARE-WE-DONE.
                            PERFORM MAIN-LOOP
                                UNTIL WE-ARE-DONE.
                            STOP RUN.

                        MAIN-LOOP.
                            DISPLAY 'SALES REP NUMBER OR ** TO END PROGRAM: '.
                            ACCEPT W-SLSREP-NUMBER.
                            IF W-SLSREP-NUMBER = '**'
                                MOVE 'YES' TO ARE-WE-DONE
                              ELSE
                                PERFORM FIND-AND-DISPLAY-SALES_REP.

                        FIND-AND-DISPLAY-SALES_REP.
                            EXEC SQL
                                SELECT LAST, FIRST, STREET, CITY, STATE, ZIP_CODE,
                                    TOTAL_COMMISSION, COMMISSION_RATE
                                INTO :W-LAST, :W-FIRST, :W-STREET, :W-CITY, :W-STATE,
                                    :W-ZIP_CODE,  :W-TOTAL-COMMISSION,
                                    :W-COMMISSION-RATE
                                FROM SALES_REP
                                WHERE SLSREP_NUMBER = :W-SLSREP-NUMBER
                            END-EXEC.
                            IF SQLCODE = 0
                                DISPLAY '       LAST NAME: ', W-LAST,
                                DISPLAY '      FIRST NAME: ', W-FIRST,
                                DISPLAY '          STREET: ', W-STREET,
                                DISPLAY '            CITY: ', W-CITY,
                                DISPLAY '           STATE: ', W-STATE,
                                DISPLAY '        ZIP CODE: ', W-ZIP-CODE
                                DISPLAY 'TOTAL COMMISSION: ', W-TOTAL-COMMISSION
                                DISPLAY ' COMMISSION RATE: ', W-COMMISSION-RATE
                              ELSE
                                DISPLAY 'THERE IS NO SUCH SALES REP'.
```

Retrieve a Single Row from a Join

You can use an embedded SQL command to join tables, as illustrated in Example 3.

EXAMPLE 3 Obtain the last name, first name, and address of the customer whose customer number is stored in the host variable, W-CUSTOMER-NUMBER, as well as the number, last name, and first name of the sales rep who represents this customer.

This query involves joining the CUSTOMER and SALES_REP tables. Because the restriction involves the primary key of the CUSTOMER table and each customer is related to exactly one sales rep, the result of the query will be a single row. The method for handling this query is similar to that for the preceding queries. The SELECT command is written as follows:

```
EXEC SQL
    SELECT CUSTOMER.LAST, CUSTOMER.FIRST, CUSTOMER.STREET,
        CUSTOMER.CITY, CUSTOMER.STATE, CUSTOMER.ZIP_CODE,
        CUSTOMER.SLSREP_NUMBER, SALES_REP.LAST,
        SALES_REP.FIRST
    INTO :W-LAST OF :W-CUSTOMER, :W-FIRST OF :W-CUSTOMER,
        :W-STREET OF :W-CUSTOMER, :W-CITY OF :W-CUSTOMER,
        :W-STATE OF :W-CUSTOMER, :W-ZIP-CODE OF :W-CUSTOMER,
        :W-SLSREP-NUMBER OF :W-CUSTOMER,
        :W-LAST OF :W-SALES-REP, :W-FIRST OF :W-SALES-REP
    FROM SALES_REP, CUSTOMER
    WHERE SALES_REP.SLSREP_NUMBER = CUSTOMER.SLSREP_NUMBER
    AND CUSTOMER.CUSTOMER_NUMBER = :W-CUSTOMER-NUMBER
END-EXEC.
```

Note: Any qualification of host variables must follow the rules of the host language rather than the rules of SQL. In COBOL, for example, this means following the name of the variable with the OF command and then by the record name. For example, to indicate the last name of a customer, the expression is :W-LAST OF :W-CUSTOMER. To indicate the last name of a sales rep, the expression is :W-LAST OF :W-SALES-REP.

Insert a Row into a Table

When you are updating SQL databases from within COBOL programs, you need to use appropriate SQL commands to update the database. For example, to add a row to a table in the database, you do not use the COBOL WRITE statement that normally is used when updating files. Instead, you use the SQL INSERT command, as illustrated in Example 4.

EXAMPLE 4 : Add a row to the SALES_REP table. The sales rep number, last name, first name, address, total commission, and credit limit already have been placed in the variables W-SLSREP-NUMBER, W-LAST, W-FIRST, W-STREET, W-CITY, W-STATE, W-ZIP-CODE, W-TOTAL-COMMISSION, and W-COMMISSION-RATE, respectively.

To insert a row into a table, use the INSERT command. The values are contained in host variables, whose names must be preceded by colons. The command is written as follows:

```
EXEC SQL
    INSERT
        INTO SALES_REP
        VALUES (:W-SLSREP-NUMBER, :W-LAST, :W-FIRST, :W-STREET,
            :W-CITY, :W-STATE, :W-ZIP-CODE, :W-TOTAL-COMMISSION,
            :W-COMMISSION-RATE)
END-EXEC.
```

The values currently in the host variables included in the INSERT statement are used to add a new row to the SALES_REP table.

Figure 8.3 shows a complete COBOL program to add sales reps. The numbered portions of the program are described in the section following the figure.

FIGURE 8.3   Program to add sales reps

```
IDENTIFICATION DIVISION.
PROGRAM-ID.      ADDSLS.

ENVIRONMENT DIVISION.

DATA DIVISION.

01  STATUS-FLAGS.
    03  ARE-WE-DONE              PIC X(3).
        88  WE-ARE-DONE                          VALUE 'YES'.
    03  IS-DATA-VALID           PIC X(3).
        88  DATA-IS-VALID                        VALUE 'YES'.

01  W-SALES-REP.
    03  W-SLSREP-NUMBER         PIC X(2).
    03  W-LAST                  PIC X(10).
    03  W-FIRST                 PIC X(8).
    03  W-STREET                PIC X(15).
    03  W-CITY                  PIC X(15).
    03  W-STATE                 PIC X(2).
    03  W-ZIP-CODE              PIC X(5).
    03  W-TOTAL-COMMISSION      PIC S9(5)V9(2).
    03  W-COMMISSION-RATE       PIC SV9(2).

    EXEC SQL
        INCLUDE DECSALES_REP
    END-EXEC.

    EXEC SQL
        INCLUDE SQLCA
    END-EXEC.

PROCEDURE-DIVISION.

MAIN-PROGRAM.
    MOVE 'NO' TO ARE-WE-DONE.
    PERFORM MAIN-LOOP
        UNTIL WE-ARE-DONE.
    STOP RUN.

MAIN-LOOP.
    DISPLAY 'SALES REP NUMBER OR ** TO END PROGRAM: '.
    ACCEPT W-SLSREP-NUMBER.
    IF W-SLSREP-NUMBER = '**'
        MOVE 'YES' TO ARE-WE-DONE
    ELSE
        PERFORM OBTAIN-REMAINING-DATA.
        PERFORM VALIDATE-DATA.
        IF DATA-IS-VALID
            PERFORM ADD-SALES_REP.

OBTAIN-REMAINING-DATA.
    DISPLAY 'ENTER LAST NAME: '.
    ACCEPT W-LAST.
    DISPLAY 'ENTER FIRST NAME: '
    ACCEPT W-FIRST.
    DISPLAY 'ENTER STREET: '.
    ACCEPT W-STREET.
    DISPLAY 'ENTER CITY: '.
    ACCEPT W-CITY.
    DISPLAY 'ENTER STATE: '.
        ACCEPT W-STATE.
        DISPLAY 'ENTER ZIP CODE: '.
        ACCEPT W-ZIP_CODE.
        DISPLAY 'ENTER TOTAL COMMISSION: '.
        ACCEPT W-TOTAL-COMMISSION.
        DISPLAY 'ENTER COMMISSION RATE: '.
        ACCEPT W-COMMISSION-RATE.

    VALIDATE-DATA.
        MOVE 'YES' TO IS-DATA-VALID.
        EXEC SQL
            SELECT LAST
                INTO :W-LAST
                FROM SALES_REP
                WHERE SLSREP_NUMBER = :W-SLSREP-NUMBER
        END-EXEC.
        IF SQLCODE = 0
            MOVE 'NO' TO IS-DATA-VALID
            DISPLAY 'ERROR - DUPLICATE SALES REP'.

    ADD-SALES_REP.
        EXEC SQL
            INSERT
            INTO SALES_REP
            VALUES (:W-SLSREP-NUMBER, :W-LAST, :W-FIRST,
                :W-STREET, :W-CITY, :W-STATE, :W-ZIP-CODE,
                :W-TOTAL-COMMISSION, :W-COMMISSION-RATE)
        END-EXEC.
```

1. This command performs the paragraph named OBTAIN-REMAINING-DATA to obtain the rest of the data concerning the sales rep from the user.

2. The paragraph named VALIDATE-DATA ensures the validity of the data entered by the user. If the data is valid, the flag named IS-DATA-VALID is set to YES. If the data is not valid, the flag is set to NO.

3. If the data is valid, the program performs the paragraph named ADD-SALES_REP to add the data to the database.

4. For each field to enter, this paragraph contains a prompt for the field (such as "ENTER LAST NAME: ") followed immediately by an ACCEPT command to obtain the desired data from the user.

5. The SELECT command determines whether there is a sales rep already in the database with the same number.

6. If SQLCODE is zero, a sales rep with the same number has been found. In this case, you should not add the sales rep. You indicate this by setting IS-DATA-VALID to NO and displaying an error message.

7. This INSERT command adds the new row to the database.

In general, the paragraph named VALIDATE-DATA is used to provide any sort of required validation. In a program to add customers, for example, this paragraph could contain logic necessary to ensure that credit limits are $750, $1,000, $1,500, or $2,000. It also could contain logic to ensure that the sales rep number entered for the customer corresponds to a sales rep number already in the database.

Change a Single Row in a Table

You used SQL commands to insert rows in a SQL database; you also need to use SQL commands to update the rows, as illustrated in Example 5.

EXAMPLE 5   Change the last name of the sales rep whose number currently is stored in W-SLSREP-NUMBER to the value currently stored in W-LAST.

Again, the only difference between this example and the update examples in Chapter 5 is the use of host variables. The statement is written as follows:

```
EXEC SQL
    UPDATE SALES_REP
         SET LAST = :W-LAST
         WHERE SLSREP_NUMBER = :W-SLSREP-NUMBER
END-EXEC.
```

Change Multiple Rows in a Table

A benefit of using SQL commands to update the database is that you can use a single command to update multiple rows—something you normally cannot accomplish in COBOL. Example 6 illustrates this capability.

EXAMPLE 6 : Add the amount stored in the host variable INCREASE-IN-RATE to the commission rate for all sales reps who currently represent any customer having a credit limit of $1,000.

To update multiple rows, the statement is written as follows:

```
EXEC SQL
    UPDATE SALES_REP
        SET COMMISSION_RATE = COMMISSION_RATE
            + :INCREASE-IN-RATE
        WHERE SLSREP_NUMBER IN
            (SELECT SLSREP_NUMBER
                FROM CUSTOMER
                WHERE CREDIT_LIMIT = 1000)
END-EXEC.
```

Delete a Single Row from a Table

Just as you would expect, if you must use SQL commands to insert and change rows in a table, you also must use SQL commands to delete rows, as illustrated in Example 7.

EXAMPLE 7 : Delete the sales rep whose number currently is stored in W-SLSREP-NUMBER from the SALES_REP table.

The formulation is written as follows:

```
EXEC SQL
    DELETE
        FROM SALES_REP
        WHERE SLSREP_NUMBER = :W-SLSREP-NUMBER
END-EXEC.
```

Delete Multiple Rows from a Table

Sometimes you need to delete more than one row from a table. If you delete an order from the ORDERS table, for example, you also need to delete all associated order lines from the ORDER_LINE table, as illustrated in Example 8.

186

EXAMPLE 8 : Delete every order line for the order whose order number currently is stored in the host variable W-ORDER-NUMBER from the ORDER_LINE table.

The formulation is written as follows:

```
EXEC SQL
     DELETE
          FROM ORDER_LINE
          WHERE ORDER_NUMBER = :W-ORDER-NUMBER
END-EXEC.
```

■ Multiple-Row SELECT

The previous examples posed no problems for COBOL because the SELECT statements retrieved only individual rows. There was an UPDATE example in which multiple rows were updated and a DELETE example in which multiple rows were deleted, but these tasks presented no difficulty. The SQL statements were executed and the updates or deletions occurred. The program could move on to the next task.

What happens when a SELECT statement produces multiple rows? What if, for example, the SELECT statement produced the number and name of every customer represented by the sales rep whose number was stored in W-SLSREP-NUMBER? Could you formulate this query as follows:

```
EXEC SQL
     SELECT CUSTOMER_NUMBER, LAST, FIRST
          INTO :W-CUSTOMER-NUMBER, :W-LAST, :W-FIRST
          FROM CUSTOMER
          WHERE SLSREP_NUMBER = :W-SLSREP-NUMBER
END-EXEC.
```

There is a problem—COBOL can process only one record at a time, whereas this SQL command produces multiple rows (records). Whose number and name is placed in W-CUSTOMER-NUMBER, W-LAST, and W-FIRST if 100 customers are retrieved? Should you make W-CUSTOMER-NUMBER, W-LAST, and W-FIRST arrays capable of holding 100 customers and, if so, what should be the size of these arrays? Fortunately, you can solve this problem by using a cursor.

Cursors

A **cursor** is a pointer to a row in the collection of rows retrieved by a SQL statement. (This is *not* the same cursor that you see on your computer screen.) The cursor advances one row at a time to provide sequential, record-at-a-time access to the retrieved rows so COBOL can process the rows. By using a cursor, COBOL can process the set of retrieved rows as though they were records in a sequential file.

To use a cursor, you must first declare it, as illustrated in Example 9.

EXAMPLE 9 : Retrieve the number, last name, and first name of every customer represented by the sales rep whose number is stored in the host variable W-SLSREP-NUMBER.

The first step in using a cursor is to declare the cursor and describe the associated query. You do this task in the WORKING-STORAGE SECTION of the DATA DIVISION. The command is written as follows:

```
EXEC SQL
    DECLARE CUSTGROUP CURSOR FOR
        SELECT CUSTOMER_NUMBER, LAST, FIRST
        FROM CUSTOMER
        WHERE SLSREP_NUMBER = :W-SLSREP-NUMBER
END-EXEC.
```

This formulation does *not* cause the query to be executed at this time; it only declares a cursor named CUSTGROUP and associates the cursor with the indicated query. Using a cursor in the PROCEDURE DIVISION involves three commands: OPEN, FETCH, and CLOSE. Opening the cursor causes the query to be executed and makes the results available to the program. Executing a fetch advances the cursor to the next row in the set of rows retrieved by the query and places the contents of the row in the indicated host variables. Finally, closing a cursor deactivates it. Data retrieved by the execution of the query is no longer available. The cursor could be opened again later and processing could begin again. If any host variables used in making the selection change, then the set of rows might change as well.

The OPEN, FETCH, and CLOSE commands used in processing a cursor are analogous to the OPEN, READ, and CLOSE commands used in processing a sequential file. Next you will examine how each of these commands is coded in COBOL.

Opening a Cursor

The formulation for the OPEN command is written as follows:

```
EXEC SQL
    OPEN CUSTGROUP
END-EXEC.
```

Figure 8.4 shows the result of opening the CUSTGROUP cursor. In the figure, assume for purposes of the example that W-SLSREP-NUMBER is set to 12 before the open statement is executed. Note that prior to the open, no rows were available. After the open, three rows are available to the program and the cursor is positioned at the first row; that is, the next FETCH command causes the contents of the first row to be placed in the indicated host variables.

FIGURE 8.4A   Before OPEN

CUSTGROUP

CUSTOMER_NUMBER	LAST	FIRST		W-CUSTOMER_NUMBER	W-LAST	W-FIRST	SQLCODE
			←next row to be fetched				0

FIGURE 8.4B   After OPEN, but before first FETCH

CUSTGROUP

CUSTOMER_NUMBER	LAST	FIRST		W-CUSTOMER_NUMBER	W-LAST	W-FIRST	SQLCODE
311	Charles	Don	←next row to be fetched				0
405	Williams	Al					
522	Nelson	Mary					

Fetching Rows from a Cursor

To fetch (get) the next row from a cursor, use the FETCH command. The FETCH command is written as follows:

```
EXEC SQL
     FETCH CUSTGROUP
          INTO :W-CUSTOMER-NUMBER, :W-LAST, :W-FIRST
END-EXEC.
```

Note that the INTO clause is associated with the FETCH command itself and not with the query used in the definition of the cursor. The execution of that query probably produces multiple rows. The execution of the FETCH command produces only a single row, so it is appropriate that the FETCH command causes data to be placed in the indicated host variables.

Figure 8.5 shows the result of four FETCH commands. Note that the first three fetches are successful. In each case, the data from the appropriate row in the cursor is placed in the indicated host variables and SQLCODE is set to zero. The fourth FETCH command is different, however, because there is no more data to fetch. In this case, the contents of the host variables are left untouched and SQLCODE is set to 100.

FIGURE 8.5A After first FETCH

CUSTGROUP

CUSTOMER_NUMBER	LAST	FIRST
311	Charles	Don
405	Williams	Al	←next row to be fetched
522	Nelson	Mary

W-CUSTOMER_NUMBER	W-LAST	W-FIRST	SQLCODE
311	Charles	Don	0

FIGURE 8.5B After second FETCH

CUSTGROUP

CUSTOMER_NUMBER	LAST	FIRST
311	Charles	Don
405	Williams	Al
522	Nelson	Mary	←next row to be fetched

W-CUSTOMER_NUMBER	W-LAST	W-FIRST	SQLCODE
405	Williams	Al	0

FIGURE 8.5C After third FETCH

CUSTGROUP

CUSTOMER_NUMBER	LAST	FIRST
311	Charles	Don
405	Williams	Al
522	Nelson	Mary
			←no more rows to be fetched

W-CUSTOMER_NUMBER	W-LAST	W-FIRST	SQLCODE
522	Nelson	Mary	0

FIGURE 8.5D   After attempting a fourth FETCH (note SQLCODE is 100)

CUSTGROUP

CUSTOMER_NUMBER	LAST	FIRST
311	Charles	Don
405	Williams	Al
522	Nelson	Mary
			←no more rows to be fetched

W-CUSTOMER_NUMBER	W-LAST	W-FIRST	SQLCODE
522	Nelson	Mary	100

Closing a Cursor

The CLOSE command is written as follows:

```
EXEC SQL
      CLOSE CUSTGROUP
END-EXEC.
```

Figure 8.6 shows the result of closing the cursor. The data is no longer available.

FIGURE 8.6   After CLOSE

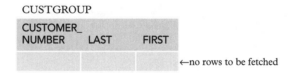

CUSTGROUP

CUSTOMER_NUMBER	LAST	FIRST
			←no rows to be fetched

Figure 8.7 shows a complete COBOL program using the cursor. Let's examine the numbered portions of the program.

FIGURE 8.7   Program to display customers of a given sales rep

```
              IDENTIFICATION DIVISION.
              PROGRAM-ID.       CUSTSLS.

              ENVIRONMENT DIVISION.

              DATA DIVISION.

              WORKING-STORAGE SECTION.

              01  STATUS-FLAGS.
                  03  ARE-WE-DONE              PIC X(3).
                      88  WE-ARE-DONE                       VALUE 'YES'.

1             01  W-CUSTOMER.
                  03  W-CUSTOMER_NUMBER    PIC X(3).
                  03  W-LAST               PIC X(10).
                  03  W-FIRST              PIC X(8).
                  03  W-STREET             PIC X(15).
                  03  W-CITY               PIC X(15).
                  03  W-STATE              PIC X(2).
                  03  W-ZIP_CODE           PIC X(5).
                  03  W-BALANCE            PIC S9(5)V9(2).
                  03  W-CREDIT-LIMIT       PIC S9(4).
                  03  W-SLSREP-NUMBER      PIC X(2).

2             EXEC SQL
                  INCLUDE DECCUSTOMER
              END-EXEC.

              EXEC SQL
                  INCLUDE SQLCA
              END-EXEC.

3             EXEC SQL
                  DECLARE CUSTGROUP CURSOR FOR
                      SELECT CUSTOMER_NUMBER, LAST, FIRST
                      FROM CUSTOMER
                      WHERE SLSREP_NUMBER = :W-SLSREP-NUMBER
              END-EXEC.

              PROCEDURE-DIVISION.

              MAIN-PROGRAM.
                  MOVE 'NO' TO ARE-WE-DONE.
                  PERFORM MAIN-LOOP
                      UNTIL WE-ARE-DONE.
                  STOP RUN.

              MAIN-LOOP.
                  DISPLAY 'SALES REP NUMBER OR ** TO END PROGRAM: '.
                  ACCEPT W-SLSREP-NUMBER.
                  IF W-SLSREP-NUMBER = '**'
                      MOVE 'YES' TO ARE-WE-DONE
                  ELSE
                      PERFORM FIND-CUSTOMERS.

              FIND-CUSTOMERS.
4             EXEC SQL
                  OPEN CUSTGROUP
                  END-EXEC.
5             PERFORM FIND-A-CUSTOMER
                      UNTIL SQLCODE = 100.
6             EXEC SQL
                  CLOSE CUSTGROUP
              END-EXEC.

              FIND-A-CUSTOMER.
7             EXEC SQL
                  FETCH CUSTGROUP
                      INTO :W-CUSTOMER_NUMBER, :W-LAST, :W-FIRST
                  END-EXEC.
8             IF SQLCODE =
                  DISPLAY 'CUSTOMER NUMBER: ', W-CUSTOMER_NUMBER
                  DISPLAY '      LAST NAME: ', W-LAST
                  DISPLAY '     FIRST NAME: ', W-FIRST.
```

1. The appropriate host variables for customer data are declared as a record named W-CUSTOMER.

2. The description of the CUSTOMER table is stored in DECCUSTOMER, so DECCUSTOMER appears in an INCLUDE statement.

3. This statement gives the declaration of the cursor (CUSTGROUP).

4. This statement opens the cursor, making the desired data available to the program.

5. This statement performs the paragraph named FIND-A-CUSTOMER until SQLCODE is set to 100. Recall that when SQLCODE is set to 100, all rows have been fetched and an attempt is made to fetch another row.

6. This statement closes the cursor.

7. This FETCH command places the information from the next row in the cursor into the indicated host variables. If no more rows exist, SQLCODE will be set to 100.

8. If SQLCODE is zero, a row was found and the data is displayed.

More Complex Cursors

The formulation of the query to define the cursor in Example 9 was simple. Any SQL query is legitimate in a cursor definition. In fact, the more complicated the requirements for retrieval, the more numerous the benefits derived by the programmer who uses embedded SQL. Consider the query in Example 10.

EXAMPLE 10 : For every order that contains an order line for the part whose part number is stored in W-PART-NUMBER, retrieve the order number, order date, order number, last name, and first name of the customer who placed the order, and the number, last name, and first name of the sales rep who represents the customer. Sort the results by customer number.

Opening and closing the cursor is done exactly as in Example 9. The only difference in the FETCH command is a different set of host variables in the INTO clause. Thus, the only real difference is the definition of the cursor itself. In this case, the cursor definition is written as follows:

```
EXEC SQL
     DECLARE ORDGROUP CURSOR FOR
          SELECT ORDERS.ORDER_NUMBER, ORDERS.ORDER_DATE,
          CUSTOMER.CUSTOMER_NUMBER, CUSTOMER.LAST, FIRST,
          SALES_REP.SLSREP_NUMBER, SALES_REP.LAST, FIRST
          FROM ORDER_LINE, ORDERS, CUSTOMER, SALES_REP
          WHERE ORDER_LINE.PART_NUMBER = :W-PART-NUMBER
          AND ORDER_LINE.ORDER_NUMBER = ORDERS.ORDER_NUMBER
          AND ORDERS.CUSTOMER_NUMBER = CUSTOMER.CUSTOMER_NUMBER
          AND CUSTOMER.SLSREP_NUMBER = SALES_REP.SLSREP_NUMBER
          ORDER BY CUSTOMER.CUSTOMER_NUMBER
END-EXEC.
```

Advantages of Cursors

The retrieval requirements in Example 10 are complicated. Yet, beyond coding the preceding cursor declaration, the programmer doesn't have to worry about the mechanics of obtaining the necessary data or placing it in the right order, because this happens automatically when the cursor is opened. To the programmer, it seems as if a sequential file already exists that has precisely the right data in it, sorted in the right order. This assumption leads to three main advantages. First, the coding in the program is greatly simplified. Second, in a normal COBOL program, the programmer must determine the most efficient way to access the data. In a program using embedded SQL, a special component of the database management system called the **optimizer** determines the best way to access the data. The programmer isn't concerned with the best way to pull the data together. In addition, if an underlying structure changes (for example, an additional index is created), the optimizer determines the best way to execute the query in view of the new structure. The program does not have to change at all. Third, if the database structure changes in such a way that the necessary information still is obtainable using a different query, the only change required in the program is the cursor definition in WORKING-STORAGE. The PROCEDURE DIVISION code is not affected.

■ Updating Cursors

You can update the rows encountered in processing cursors. In order to indicate that an update is required, you include an additional clause—FOR UPDATE OF—in the cursor definition. For example, consider the update requirement in Example 11.

EXAMPLE 11 : Add $100 to the credit limit for every customer represented by the sales rep whose number currently is stored in the host variable W-SLSREP-NUMBER, whose balance is not over the credit limit, and whose credit limit is $500 or less. Add $200 to the credit limit of every customer of this sales rep whose balance is not over the credit limit and whose credit limit is more than $500. Write the number, last name, and first name of every customer of this sales rep whose balance is greater than the credit limit.

194

The cursor declaration is written as follows:

```
EXEC SQL
     DECLARE CREDGROUP CURSOR FOR
          SELECT CUSTOMER_NUMBER, LAST, FIRST, BALANCE,
          CREDIT_LIMIT
          FROM CUSTOMER
          WHERE SLSREP_NUMBER = :W-SLSREP-NUMBER
          FOR UPDATE OF CREDIT_LIMIT
END-EXEC.
```

To update the credit limits, you must include the FOR UPDATE OF CREDIT_LIMIT clause in the cursor declaration. The PROCEDURE DIVISION code for the OPEN and CLOSE statements is the same one you used previously. The code to fetch a row, determine whether it was fetched, and then take appropriate action is written as follows:

```
EXEC SQL
     FETCH CREDGROUP
          INTO :W-CUSTOMER-NUMBER, :W-LAST, :W-FIRST,
          :W-BALANCE, :W-CREDIT-LIMIT
END-EXEC.
IF SQLCODE = 100
     MOVE "NO" TO ARE-THERE-MORE-CUSTOMERS
ELSE
     PERFORM CUSTOMER-UPDATE.
CUSTOMER-UPDATE.
     IF W-CREDIT-LIMIT > W-BALANCE
          DISPLAY W-FIRST, W-LAST
     ELSE IF W-CREDIT-LIMIT > 500
          EXEC SQL
               UPDATE CUSTOMER
                    SET CREDIT_LIMIT = CREDIT_LIMIT + 200
                    WHERE CURRENT OF CREDGROUP
          END-EXEC
     ELSE
          EXEC SQL
               UPDATE CUSTOMER
                    SET CREDIT_LIMIT = CREDIT_LIMIT + 100
                    WHERE CURRENT OF CREDGROUP
END-EXEC.
```

The preceding code is in a loop that is performed until the flag ARE-THERE-MORE-CUSTOMERS is set to NO. The FETCH command either makes the next retrieved row available to the program with the values placed in the variables W-CUSTOMER-NUMBER, W-LAST, W-FIRST, W-BALANCE, and W-CREDIT-LIMIT, or it sets SQLCODE to 100 to indicate that no more rows were retrieved. The code that comes after the FETCH command sets the flag to NO if SQLCODE is set to 100. If SQLCODE is not set to 100, then the update routine is performed.

In the update routine, the credit limit first is compared with the balance. If the balance is larger, then a message is printed. If the balance is not larger, then the credit limit is compared with $500. If the credit limit is larger, then the row is updated by adding $200 to the credit limit.

If the credit limit is not larger, then the row is updated by adding $100 to the credit limit. The WHERE CURRENT OF CREDGROUP clause indicates that the update is to apply only to the row just fetched. Without this clause, and in the absence of any WHERE clause to restrict the scope of the update, this command would update *every* customer's credit limit.

■ Error Handling

Programs must be able to handle exceptional conditions that can arise when the database is accessed. Because problems are communicated through a value in SQLCODE, one legitimate way to handle problems is to check the value in SQLCODE after each executable SQL statement and then to take appropriate action based on the indicated problem. However, with all the potential conditions, this method becomes very cumbersome. Fortunately, as you will see, there is another way to handle such conditions, by using the WHENEVER statement.

There are two types of error conditions that need to be addressed. The first type consists of unusual but normal conditions, such as not retrieving any data to match a given condition, attempting to store a row that violates a duplicates clause, and so on. The value in SQLCODE for such conditions is a positive number. The appropriate action is to print an error message and continue processing. The appropriate action for the END OF DATA (SQLCODE-100) message probably is termination of some loop and continuation with the rest of the program. No error message is required.

The other type of condition is far more serious because it consists of the abnormal and unexpected conditions or, in a very real sense, the fatal ones. Examples of this type include lack of space in the database, a damaged database, and so on. The value in SQLCODE for such conditions is a negative number. The appropriate action usually is to print a final message that indicates the problem that occurred, and then terminate the program.

You can use the WHENEVER statement to handle these errors in a global way. The following example of the WHENEVER statement illustrates a typical way of handling these conditions in a program. The WHENEVER statement is written as follows:

```
EXEC SQL
     WHENEVER SQLERROR     GOTO ERROR-PROCESSING-ROUTINE
END-EXEC.
EXEC SQL
     WHENEVER SQLWARNING CONTINUE
END-EXEC.
EXEC SQL
     WHENEVER NOT FOUND CONTINUE
END-EXEC.
```

In the WHENEVER statement, SQLERROR represents the abnormal or fatal conditions (SQLCODE< 0), SQLWARNING represents the unusual but normal conditions (SQLCODE> 0), and NOT FOUND represents the special warning END OF DATA (SQLCODE=100). The WHENEVER statement ends either with GOTO followed by a section or paragraph name or with the word CONTINUE. The WHENEVER statement indicates how each of these conditions should be handled if and when they occur.

In the preceding WHENEVER statements, if a fatal condition occurs, the program is to proceed immediately to a paragraph (or section) named ERROR-PROCESSING-ROUTINE. This paragraph probably has been constructed by the organization and will be the same in each program that uses embedded SQL. Typically, this paragraph contains statements to display the SQLCODE together with any other useful information that identifies the problem, followed by a STOP RUN command. With this paragraph in place, the remainder of the program does not have to check continually for all possible errors of this type.

If an unusual but normal condition arises, or if the special NOT FOUND condition arises, however, processing should continue without any special action. This means that appropriate tests of SQLCODE must be included at appropriate places. While these tests also can be accomplished using the WHENEVER clause, doing the testing yourself provides a cleaner structure for the program. The built-in GOTO statement of the WHENEVER clause can cause problems even in well-structured programs.

In this chapter, you learned how to embed SQL commands in a procedural language. You learned how to distinguish between column names in SQL and COBOL host variables. You used embedded SQL commands to retrieve single rows, insert new rows, change existing rows, and delete rows. Then you examined the difficulties associated with embedded SQL commands that retrieved multiple rows and saw how to use cursors to address these difficulties and update a database. Finally, you learned how to handle errors in COBOL programs that contain embedded SQL commands.

■ SUMMARY

1. To embed SQL commands in a COBOL program, precede the SQL command with EXEC SQL, and follow the command with END-EXEC.

2. Statements to define the tables to be accessed must appear in the DATA DIVISION.

3. The DATA DIVISION must contain the INCLUDE SQLCA statement, which allows access to the SQL communication area.

4. You can use host language variables (variables that are not columns within a table) in embedded SQL commands by preceding the variable name with a colon.

5. You can use SELECT statements as embedded SQL commands in COBOL programs only when a single row is retrieved.

6. To place the results of a SELECT statement into host language variables, use the INTO clause in the SELECT command.

7. You can use INSERT, UPDATE, and DELETE statements in COBOL programs, even when they affect more than one row.

8. If a SELECT statement is to retrieve more than one row, it must be used to define a cursor that will be used to select one row at a time.

9. To activate a cursor, use the OPEN command to execute the query in the cursor definition.

10. To select the next row in COBOL, use the FETCH command.

11. To deactivate a cursor, use the CLOSE command. The rows initially retrieved will no longer be available to COBOL.

12. Data in the tables on which a cursor is based can be updated by including the WHERE CURRENT OF cursor name clause in the update statement. This clause updates only the current (most recently fetched) row.

13. To see if an error has occurred, examine the value in SQLCODE.

14. Rather than checking SQLCODE in every place in the program where errors could occur, use the WHENEVER clause.

▮ EXERCISES (Premiere Products)

Note: Your instructor may substitute another language for COBOL in the following exercises.

1. Assuming that the appropriate entries have been made in the DATA DIVISION of a COBOL program, give the PROCEDURE DIVISION code for each of the following:

   a. Obtain the description and unit price of the part whose part number currently is stored in W-PART-NUMBER. Place these values in the variables W-PART-DESCRIPTION and W-UNIT-PRICE, respectively.

   b. Obtain the order date, customer number, and name for the order whose number currently is stored in W-ORDER-NUMBER. Place these values in the variables W-ORDER-DATE, W-CUSTOMER-NUMBER, W-LAST, and W-FIRST, respectively.

   c. Add a row to the PART table. The data currently is stored in the columns within the W-PART record.

   d. Change the description of the part whose number is stored in W-PART-NUMBER to the value currently found in W-PART-DESCRIPTION.

   e. Increase the price of every part in item class HW by 5%.

   f. Delete the part whose number is stored in W-PART-NUMBER.

2. Retrieve the part number, part description, item class, and unit price of every part located in the warehouse whose number is stored in W-WAREHOUSE-NUMBER. In addition, you need to be able to update the unit price.

   a. Write an appropriate cursor description.

b. Write all statements that will be included in the PROCEDURE DIVISION and that relate to processing the database through this cursor.

c. Write the additional PROCEDURE DIVISION code that will update every part that is in item class HW by adding 5% to the unit price and every part in item class SG by adding 10% to the unit price. (You must use the cursor in your answer.)

3. If you have access to embedded SQL, write and run the programs created in Exercises 1 and 2.

■EXERCISES (Henry Books)

Note: Your instructor may substitute another language for COBOL in the following exercises.

1. Assuming that the appropriate entries have been made in the DATA DIVISION of a COBOL program, give the PROCEDURE DIVISION code for each of the following:

a. Obtain the name and city of the publisher whose publisher code currently is stored in W-PUBLISHER-CODE. Place these values in the variables W-PUBLISHER-NAME and W-PUBLISHER-CITY, respectively.

b. Obtain the book title, publisher code, and publisher name for every book whose code currently is stored in W-BOOK-CODE. Place these values in the variables W-BOOK-TITLE, W-PUBLISHER-CODE, and W-PUBLISHER-NAME, respectively.

c. Add a row to the BRANCH table. The data currently is stored in the fields within the W-BRANCH record.

d. Change the title of the book whose code is stored in W-BOOK-CODE to the value currently found in W-BOOK-TITLE.

e. Increase the price by 3% for every book whose publisher code is BB.

f. Delete the book whose code is stored in W-BOOK-CODE.

2. Retrieve the book code, book title, publisher code, and book price for every book whose book type is stored in W-BOOK-TYPE. In addition, you need to be able to update the book price.

a. Write an appropriate cursor description.

b. Write all statements that will be included in the PROCEDURE DIVISION and that relate to processing the database through this cursor.

c. Write the additional PROCEDURE DIVISION code that will increase the book price by 4% for every book whose type is PSY, and by 3% for every book whose type is SUS. (*Hint:* You must use the cursor in your answer.)

3. If you have access to embedded SQL, write and run the programs created in Exercises 1 and 2.

APPENDIX

SQL Reference

You can use this appendix to obtain details concerning important components and syntax of the SQL language. Items are arranged alphabetically. Each item contains a description and, where appropriate, both an example and a description of the query results. Some SQL commands also include a description of all the clauses associated with them. For each clause, there is a brief description of the clause and an indication of whether the clause is required or optional.

Aliases

You can specify an alias (alternative name) for each table in a query. You can use the alias in the rest of the command by following the name of the table with a space and the alias name.

The following command creates an alias named S for the SALES_REP table and an alias named C for the CUSTOMER table:

```
SELECT S.SLSREP_NUMBER, S.LAST, S.FIRST, C.CUSTOMER_NUMBER,
C.LAST, C.FIRST
FROM SALES_REP S, CUSTOMER C
WHERE S.SLSREP_NUMBER = C.SLSREP_NUMBER;
```

ALTER TABLE

Use the ALTER TABLE command to change the structure of a table. As shown in Table A-1, you type the ALTER TABLE command, followed by the table name, and then the alteration to perform.

TABLE A-1

ALTER TABLE
command

Clause	Description	Required?
ALTER TABLE table name	Indicates name of table to be altered.	Yes
alteration	Indicates type of alteration to be performed.	Yes

The following command alters the CUSTOMER table by adding a new CUSTOMER_TYPE column:

```
ALTER TABLE CUSTOMER
ADD CUSTOMER_TYPE CHAR(1);
```

The following command changes the CITY column in the CUSTOMER table so that it cannot accept nulls:

```
ALTER TABLE CUSTOMER
MODIFY CITY NOT NULL;
```

Column or Expression List (SELECT Clause)

To select columns, use the SELECT clause followed by the list of columns, separated by commas.

The following SELECT clause selects the CUSTOMER_NUMBER, LAST, FIRST, and BALANCE columns:

```
SELECT CUSTOMER_NUMBER, LAST, FIRST, BALANCE
```

Use an asterisk in a SELECT clause to select all columns in a table.

The following SELECT clause selects all columns:

```
SELECT *
```

Computed Columns

You can use a computation in place of a column by typing the computation. For readability, you can type the computation in parentheses, even though it is not necessary to do so.

The following SELECT clause selects the CUSTOMER_NUMBER, LAST, and FIRST columns as well as the results of subtracting the BALANCE column from the CREDIT_LIMIT column:

```
SELECT CUSTOMER_NUMBER, LAST, FIRST, (CREDIT_LIMIT - BALANCE)
```

The DISTINCT Operator

To avoid selecting duplicate values in a command, use the DISTINCT operator. If you omit the DISTINCT operator in the command and a customer number appears on multiple rows in the ORDERS table, that customer number will appear on multiple rows in the query results.

The following query selects all customer numbers from the ORDERS table, but it lists each customer number only once in the results:

```
SELECT DISTINCT(CUSTOMER_NUMBER)
FROM ORDERS;
```

Functions

You can use functions in a SELECT clause. The most commonly used functions are AVG (to calculate an average), COUNT (to count the number of rows), MAX (to determine the maximum value), MIN (to determine the minimum value), and SUM (to calculate a total).

The following SELECT clause calculates the average balance:

```
SELECT AVG(BALANCE)
```

■ COMMIT

Use the COMMIT command to make any updates since the last command permanent. If no previous COMMIT command has been executed, the COMMIT command will make all the updates during the current work session permanent immediately. All updates become permanent automatically when you exit SQL. Table A-2 describes the COMMIT command.

Clause	Description	Required?
COMMIT	Indicates that a COMMIT is to be performed.	Yes

The following command makes all updates since the most recent COMMIT command permanent:

```
COMMIT;
```

■ Conditions

A condition is an expression that evaluates to either true or false. When you use a condition in a WHERE clause, the results of the query contain those rows for which the condition is true. You can create simple conditions and compound conditions using the BETWEEN, LIKE, IN, EXISTS, ALL, and ANY operators, as described in the following sections.

Simple Conditions

A simple condition has the form: column name, comparison operator, and then either another column name or a value. The available comparison operators are = (equal to), < (less than), > (greater than), <= (less than or equal to), >= (greater than or equal to), and < > (not equal to).

The following WHERE clause uses a condition to select rows where the balance is greater than the credit limit:

```
WHERE BALANCE > CREDIT_LIMIT
```

Compound Conditions

Compound conditions are formed by connecting two or more simple conditions using the AND, OR, and NOT operators. When simple conditions are connected by the AND operator, all of the simple conditions must be true in order for the compound condition to be true. When simple conditions are connected by the OR operator, the compound condition will be true whenever any one of the simple conditions is true. Preceding a condition by the NOT operator reverses the truth of the original condition.

The following WHERE clause is true if the warehouse number is equal to 3 *or* the units on hand is greater than 100, *or* both:

```
WHERE (WAREHOUSE_NUMBER = '3') OR (UNITS_ON_HAND > 100)
```

The following WHERE clause is true if the warehouse number is equal to 3 *and* the units on hand is greater than 100:

```
WHERE (WAREHOUSE_NUMBER = '3') AND (UNITS_ON_HAND > 100)
```

The following WHERE clause is true if the warehouse number is *not* equal to 3:

```
WHERE NOT (WAREHOUSE_NUMBER = '3')
```

BETWEEN Conditions

You can use BETWEEN to determine if a value is within a range of values.

The following WHERE clause is true if the balance is between 500 and 1,000:

```
WHERE BALANCE BETWEEN 500 AND 1000
```

LIKE Conditions

LIKE conditions use wildcards to select rows. Use the percent (%) wildcard to represent any collection of characters. The condition LIKE '%Pine%' will be true for data consisting of any character or characters, followed by the letters "Pine," followed by any other character or characters. Another wildcard symbol is the underscore (_), which represents any individual character. For example, "T_m" represents the letter "T," followed by any single character, followed by the letter "m," and would be true for a collection of characters such as Tim, Tom, or T3m.

The following WHERE clause is true if the value in the STREET column is Pine, Pinell, Pines, or any other value that contains "Pine:"

```
WHERE STREET LIKE '%Pine%'
```

IN Conditions

You can use IN to determine if a value is in some specific collection of values.

The following WHERE clause is true if the credit limit is 1,000, 1,500, or 2,000:

```
WHERE CREDIT_LIMIT IN (1000, 1500, 2000)
```

The following WHERE clause is true if the part number is in the collection of part numbers associated with order number 12491:

```
WHERE PART_NUMBER IN
(SELECT PART_NUMBER
FROM ORDER_LINE
WHERE ORDER_NUMBER = '12491')
```

EXISTS Conditions

You can use EXISTS to determine if the results of a subquery contain at least one row.

The following WHERE clause is true if the results of the subquery contain at least one row, that is, there is at least one order line with the desired order number and on which the part number is BT04:

```
WHERE EXISTS
(SELECT *
FROM ORDER_LINE
WHERE ORDERS.ORDER_NUMBER = ORDER_LINE.ORDER_NUMBER
AND PART_NUMBER = 'BT04')
```

ALL and ANY

You can use ALL or ANY with subqueries. If you precede the subquery by ALL, the condition is true only if it is satisfied for all values produced by the subquery. If you precede the subquery by ANY, the condition is true if it is satisfied for any value (one or more) produced by the subquery.

The following WHERE clause is true if the balance is greater than every balance contained in the results of the subquery:

```
WHERE BALANCE > ALL
(SELECT BALANCE
FROM CUSTOMER
WHERE SLSREP_NUMBER = '12')
```

The following WHERE clause is true if the balance is greater than at least one balance contained in the results of the subquery:

```
WHERE BALANCE > ANY
(SELECT BALANCE
FROM CUSTOMER
WHERE SLSREP_NUMBER = '12')
```

■ CREATE INDEX

Use the CREATE INDEX command to create an index for a table. Table A-3 describes the CREATE INDEX command.

TABLE A-3
· · · · · · · · · · · · · · · ·
CREATE INDEX
command

Clause	Description	Required?
CREATE INDEX index name	Indicates the name of the index.	Yes
ON table name	Indicates the table for which the index is to be created.	Yes
column list	Indicates the column or columns on which the index is to be based.	Yes

The following CREATE INDEX command creates an index named CUSTNAME for the CUSTOMER table on the combination of the LAST and FIRST columns:

```
CREATE INDEX CUSTNAME
ON CUSTOMER
(LAST, FIRST);
```

■ CREATE TABLE

Use the CREATE TABLE command to define the structure of a new table. Table A-4 describes the CREATE TABLE command.

Clause	Description	Required?
CREATE TABLE table name	Indicates the name of the table to be created.	Yes
(column and data type list)	Indicates the columns that comprise the table along with their corresponding data types (see Data Types section).	Yes

The following CREATE TABLE command creates the SALES_REP table and its associated columns and data types:

```
CREATE TABLE SALES_REP (SLSREP_NUMBER CHAR(2),
LAST CHAR(10),
FIRST CHAR(8),
STREET CHAR(15),
CITY CHAR(15),
STATE CHAR(2),
ZIP_CODE CHAR(5),
TOTAL_COMMISSION DECIMAL(7,2),
COMMISSION_RATE DECIMAL(3,2) );
```

■ CREATE VIEW

Use the CREATE VIEW command to create a view. Table A-5 describes the CREATE VIEW command.

Clause	Description	Required?
CREATE VIEW view name AS	Indicates the name of the view to be created.	Yes
query	Indicates the defining query for the view.	Yes

The following CREATE VIEW command creates a view named HOUSEWARES, which consists of the part number, part description, units on hand, and unit price for all rows in the PART table on which the ITEM class is HW:

```
CREATE VIEW HOUSEWARES AS
SELECT PART_NUMBER, PART_DESCRIPTION, UNITS_ON_HAND, UNIT_PRICE
FROM PART
WHERE ITEM_CLASS = 'HW';
```

■ Data Types

Table A-6 describes the data types that you can use in a CREATE TABLE command.

Data Type	Description
CHAR (n)	Character string *n* characters long. Columns that contain numbers but will not be used for arithmetic operations usually are assigned a data type of CHAR. The CUSTOMER_NUMBER column, for example, is a CHAR column because the customer numbers will not be used in any calculations.
DATE	Dates in the form DD-MON-YYYY or MM/DD/YYYY. For example, May 12, 2002 could be stored as 12-MAY-2002 or as 5/12/2002. (**Note:** The specific format in which the dates are stored varies from one implementation of SQL to another.)
DECIMAL (p,q)	Decimal number *p* digits long with *q* of these being decimal places to the right of the decimal point. The data type DECIMAL(5,2) represents a number with three places to the left of the decimal and two places to the right (for example, 100.00). (**Note:** The specific meaning of DECIMAL varies from one implementation of SQL to another. In some implementations, the decimal point counts as one of the places and in other implementations it does not. Likewise, in some implementations a minus sign counts as one of the places, but in others it does not.)
INTEGER	Integers (numbers without a decimal part); the acceptable range is -2147483648 to 2147483647.
SMALLINT	Like INTEGER but does not occupy as much space; range is -32768 to 32767. This data type is a better choice than INTEGER if you are certain that numbers will be within the indicated range.

■ DELETE Rows

Use the DELETE command to delete one or more rows from a table. Table A-7 describes the DELETE command.

Clause	Description	Required?
DELETE table name	Indicates the table from which the row or rows are to be deleted.	Yes
WHERE condition	Indicates a condition. Those rows for which the condition is true will be retrieved and deleted.	No (If you omit the WHERE clause, all rows will be deleted.)

The following DELETE command deletes any row from the CUSTOMER table on which the last name is Williams:

```
DELETE CUSTOMER
WHERE LAST = 'Williams';
```

■ DROP INDEX

Use the DROP INDEX command to delete an index, as shown in Table A-8.

Clause	Description	Required?
DROP INDEX index name	Indicates the name of the index to be dropped.	Yes

The following DROP INDEX command deletes the index named CREDNAME:

```
DROP INDEX CREDNAME;
```

■ DROP TABLE

Use the DROP TABLE command to delete a table, as shown in Table A-9.

Clause	Description	Required?
DROP TABLE table name	Indicates name of table to be dropped.	Yes

The following DROP TABLE command deletes the table named SALES_REP:

```
DROP TABLE SALES_REP;
```

■ DROP VIEW

Use the DROP VIEW command to delete a view, as shown in Table A-10.

Clause	Description	Required?
DROP VIEW view name	Indicates the name of the view to be dropped.	Yes

The following DROP VIEW command deletes the view named SLSREP:

```
DROP VIEW SLSREP;
```

■ GRANT

Use the GRANT command to grant privileges to a user. Table A-11 describes the GRANT command.

Clause	Description	Required?
GRANT privilege	Indicates the type of privilege(s) to be granted.	Yes
ON database object	Indicates the database object(s) to which the privilege(s) pertain.	Yes
TO user name	Indicates the user(s) to whom the privilege(s) are to be granted. To grant the privilege(s) to all users, use the TO PUBLIC clause.	Yes

The following GRANT command grants the user named Jones the privilege of selecting rows from the SALES_REP table:

```
GRANT SELECT
ON SALES_REP
TO JONES;
```

■ INSERT INTO (Query)

Use the INSERT INTO command with a query to insert the rows retrieved by a query into a table. As shown in Table A-12, you must indicate the name of the table into which the row(s) will be inserted and the query whose results will be inserted into the named table.

TABLE A-12
· · · · · · · · · · · · · · · · · ·
INSERT INTO
(query) command

Clause	Description	Required?
INSERT INTO table name	Indicates the name of the table into which the row(s) will be inserted.	Yes
query	Indicates the query whose results will be inserted into the table.	Yes

The following INSERT INTO command creates a table named SMALL_CUST and inserts the rows selected by the query into it:

```
INSERT INTO SMALL_CUST
SELECT *
FROM CUSTOMER
WHERE CREDIT_LIMIT <= 1200;
```

■ INSERT INTO (Values)

Use the INSERT INTO command and the VALUES clause to insert a row into a table by specifying the values for each of the columns. As shown in Table A-13, you must indicate the table into which to insert the values, and then list the values to insert in parentheses.

TABLE A-13
· · · · · · · · · · · · · · · · · ·
INSERT INTO
(values)
command

Clause	Description	Required?
INSERT INTO table name	Indicates the name of the table into which the row will be inserted.	Yes
VALUES (values list)	Indicates the values for each of the columns on the new row.	Yes

The following INSERT INTO command inserts the values shown in parentheses as a new row in the CUSTOMER table:

```
INSERT INTO CUSTOMER
VALUES
('124','Adams','Sally','481 Oak','Lansing','MI','49224',818.75,
1000,'03');
```

■ Integrity

You can use the ALTER TABLE command with an appropriate CHECK, ADD PRIMARY KEY, or ADD FOREIGN KEY clause to specify integrity. Table A-14 describes the ALTER TABLE command for specifying integrity.

TABLE A-14
....
Integrity options

Clause	Description	Required?
ALTER TABLE table name	Indicates the table for which integrity is being specified.	Yes
integrity clause	CHECK, ADD PRIMARY KEY, or ADD FOREIGN KEY	Yes

The following ALTER TABLE command changes the PART table so that the only legal values for the ITEM_CLASS column are AP, HW, and SG:

```
ALTER TABLE PART
CHECK (ITEM_CLASS IN ('AP','HW','SG') );
```

The following ALTER TABLE command changes the SALES_REP table so that the SLSREP_NUMBER column is the table's primary key:

```
ALTER TABLE SALES_REP
ADD PRIMARY KEY(SLSREP_NUMBER);
```

The following ALTER TABLE command changes the CUSTOMER table so that the SLSREP_NUMBER column in the CUSTOMER table is a foreign key referencing the primary key of the SALES_REP table:

```
ALTER TABLE CUSTOMER
ADD FOREIGN KEY(SLSREP_NUMBER) REFERENCES SALES_REP;
```

■ REVOKE

Use the REVOKE command to revoke privileges from a user. Table A-15 describes the REVOKE command.

Clause	Description	Required?
REVOKE privilege	Indicates the type of privilege(s) to be revoked.	Yes
ON database object	Indicates the database object(s) to which the privilege pertains.	Yes
FROM user name	Indicates the user name(s) from whom the privilege(s) are to be revoked.	Yes

The following REVOKE command revokes the SELECT privilege for the SALES_REP table from the user named Jones:

```
REVOKE SELECT
ON SALES_REP
FROM JONES;
```

ROLLBACK

Use the ROLLBACK command to reverse (undo) all updates since the execution of the previous COMMIT command. If no COMMIT command has been executed, the command will undo all changes during the current work session. Table A-16 describes the ROLLBACK command.

TABLE A-16

ROLLBACK
command

Clause	Description	Required?
ROLLBACK	Indicates that a rollback is to be performed.	Yes

The following command reverses all updates made since the time of the last COMMIT command:

```
ROLLBACK;
```

SELECT

Use the SELECT command to retrieve data from a table or from multiple tables. Table A-17 describes the SELECT command.

Clause	Description	Required?
SELECT column or expression list	Indicates the column(s) and/or expression(s) to be retrieved.	Yes
FROM table list	Indicates the table(s) required for the query.	Yes
WHERE condition	Indicates one or more conditions. Only the rows for which the condition(s) are true will be retrieved.	No (If you omit the WHERE clause, all rows will be retrieved.)
GROUP BY column list	Indicates column(s) on which rows are to be grouped.	No (If you omit the GROUP BY clause, no grouping will occur.)
HAVING condition involving groups	Indicates a condition for groups. Only groups for which the condition is true will be included in query results. You use the HAVING clause only if the query output is grouped.	No (If you omit the HAVING clause, all groups will be included.)
ORDER BY column or expression list	Indicates column(s) on which the query output is to be sorted.	No (If you omit the ORDER BY clause, no sorting will occur.)

The following SELECT command selects the customer number, order number, order date, and the sum of the product of the number ordered and unit price:

```
SELECT CUSTOMER_NUMBER, ORDERS.ORDER_NUMBER, ORDER_DATE,
SUM(NUMBER_ORDERED * QUOTED_PRICE)
FROM ORDERS, ORDER_LINE
WHERE ORDERS.ORDER_NUMBER = ORDER_LINE.ORDER_NUMBER
GROUP BY ORDERS.ORDER_NUMBER, CUSTOMER_NUMBER, ORDER_DATE
HAVING SUM(NUMBER_ORDERED * QUOTED_PRICE) > 100
ORDER BY ORDERS.ORDER_NUMBER;
```

The data comes from the ORDERS and ORDER_LINE tables. Only rows on which the order numbers are the same are included. The data is grouped by order number, customer number, and date. Only those groups on which the sum of the product of number ordered and the quoted price is greater than 100 will be displayed. The results are ordered by order number.

◼ Subqueries

You can use one query within another. The inner query is called a subquery and it is evaluated first. The outer query is evaluated next, producing the part description for each part whose part number is in the list.

The following command contains a subquery that produces a list of part numbers included in order number 12491:

```
SELECT PART_DESCRIPTION
FROM PART
WHERE PART_NUMBER IN
(SELECT PART_NUMBER
FROM ORDER_LINE
WHERE ORDER_NUMBER = '12491');
```

◼ UNION, INTERSECT, and MINUS

Connecting two SELECT commands with UNION produces all the rows that would be in the results of the first command, the second command, or both. Connecting two SELECT commands with INTERSECT produces all the rows that would be in the results of both commands. Connecting two SELECT commands with MINUS produces all the rows that would be in the results of the first command, but *not* in the results of the second command. Table A-18 describes the UNION, INTERSECT, and MINUS operators.

TABLE A-18

⋯⋯⋯⋯⋯⋯⋯⋯
UNION, INTERSECT, and MINUS operators

Operator	Description
UNION	Produces all the rows that would be in the results of the first query, the second query, or both.
INTERSECT	Produces all the rows that would be in the results of both queries.
MINUS	Produces all the rows that would be in the results of the first query but not in the results of the second query.

The following query displays the customer number, last name, and first name of all customers who are represented by sales rep 12, *or* who have orders, *or* both:

```
SELECT CUSTOMER_NUMBER, LAST, FIRST
FROM CUSTOMER
WHERE SLSREP_NUMBER = '12'
UNION
SELECT CUSTOMER.CUSTOMER_NUMBER, LAST, FIRST
FROM CUSTOMER, ORDERS
WHERE CUSTOMER.CUSTOMER_NUMBER = ORDERS.CUSTOMER_NUMBER;
```

The following query displays the customer number, last name, and first name of all customers who are represented by sales rep 12 *and* who have orders:

```
SELECT CUSTOMER_NUMBER, LAST, FIRST
FROM CUSTOMER
WHERE SLSREP_NUMBER = '12'
INTERSECT
SELECT CUSTOMER.CUSTOMER_NUMBER, LAST, FIRST
FROM CUSTOMER, ORDERS
WHERE CUSTOMER.CUSTOMER_NUMBER = ORDERS.CUSTOMER_NUMBER;
```

The following query displays the customer number, last name, and first name of all customers who are represented by sales rep 12 but who do *not* have orders:

```
SELECT CUSTOMER_NUMBER, LAST, FIRST
FROM CUSTOMER
WHERE SLSREP_NUMBER = '12'
MINUS
SELECT CUSTOMER.CUSTOMER_NUMBER, LAST, FIRST
FROM CUSTOMER, ORDERS
WHERE CUSTOMER.CUSTOMER_NUMBER = ORDERS.CUSTOMER_NUMBER;
```

■ UPDATE

Use the UPDATE command to change the contents of one or more rows in a table. Table A-19 describes the UPDATE command.

	Clause	Description	Required?
TABLE A-19 · · · · · · · · · · · · · · · UPDATE command	UPDATE table name	Indicates the table whose contents will be changed.	Yes
	SET column = expression	Indicates the column to be changed, along with an expression that provides the new value.	Yes
	WHERE condition	Indicates a condition. The change will occur on only those rows for which the condition is true.	No (If you omit the WHERE clause, all rows will be updated.)

The following UPDATE command changes the last name on the row on which the customer number is 256 to Jones:

```
UPDATE CUSTOMER
SET LAST = 'Jones'
WHERE CUSTOMER_NUMBER = '256';
```

APPENDIX

"How do I" Reference

This appendix answers frequently asked questions about how to accomplish a variety of tasks using SQL. Use the second column to locate the correct section in Appendix A that answers your question.

How do I?	Review The Named Section(s) in Appendix A
Add columns to an existing table?	ALTER TABLE
Add rows?	INSERT INTO (Values)
Calculate a statistic (sum, average, maximum, minimum, or count)?	1. SELECT 2. COLUMN OR EXPRESSION LIST (SELECT CLAUSE) (Use the appropriate function in the query.)
Change rows?	UPDATE
Create a data type for a column?	1. DATA TYPES 2. CREATE TABLE
Create a table?	CREATE TABLE
Create a view?	CREATE VIEW
Create an index?	CREATE INDEX
Delete a table?	DROP TABLE
Delete a view?	DROP VIEW
Delete an index?	DROP INDEX
Delete rows?	DELETE ROWS
Drop a table?	DROP TABLE
Drop a view?	DROP VIEW
Drop an index?	DROP INDEX
Grant a privilege?	GRANT
Group data in a query?	SELECT (Use a GROUP BY clause.)
Insert rows using a query?	INSERT INTO (Query)
Insert rows?	INSERT INTO (Values)
Join tables?	CONDITIONS (Include a WHERE clause to relate the tables.)
Make updates permanent?	COMMIT
Order query results?	SELECT (Use the ORDER BY clause.)
Prohibit nulls?	1. CREATE TABLE 2. ALTER TABLE (Include the NOT NULL clause in a CREATE TABLE or ALTER TABLE command.)
Remove a privilege?	REVOKE
Remove rows?	DELETE ROWS

How do I?	Review The Named Section(s) in Appendix A (continued)
Retrieve all columns?	1. SELECT 2. COLUMN OR EXPRESSION LIST (SELECT CLAUSE) (Type * in the SELECT clause.)
Retrieve all rows?	SELECT (Omit the WHERE clause.)
Retrieve only certain columns?	1. SELECT 2. COLUMN OR EXPRESSION LIST (SELECT CLAUSES) (Type the list of columns in the SELECT clause.)
Revoke a privilege?	REVOKE
Select all columns?	1. SELECT 2. COLUMN OR EXPRESSION LIST (SELECT CLAUSE) (Type * in the SELECT clause.)
Select all rows?	SELECT (Omit the WHERE clause.)
Select only certain columns?	1. SELECT 2. COLUMN OR EXPRESSION LIST (SELECT CLAUSE) (Type the list of columns in the SELECT clause.)
Select only certain rows?	1. SELECT 2. CONDITIONS (Use a WHERE clause.)
Sort query results?	SELECT (Use an ORDER BY clause.)
Specify a foreign key?	INTEGRITY (Use the ADD FOREIGN KEY clause in an ALTER TABLE command.)
Specify a primary key?	INTEGRITY (Use the ADD PRIMARY KEY clause in an ALTER TABLE command.)
Specify a privilege?	GRANT
Specify integrity?	INTEGRITY (Use a CHECK, ADD PRIMARY KEY, and/or FOREIGN KEY clause in an ALTER TABLE command.)
Specify legal values?	INTEGRITY (Use the CHECK clause in an ALTER TABLE command.)
Undo updates?	ROLLBACK
Update rows?	UPDATE
Use a calculated field?	1. SELECT 2. COLUMN OR EXPRESSION LIST (SELECT CLAUSE) (Enter a calculation in the query.)
Use a compound condition in a query?	CONDITIONS

How do I?	Review The Named Section(s) in Appendix A (continued)
Use a compound condition?	1. SELECT 2. CONDITIONS (Use simple conditions connected by AND, OR, or NOT in a WHERE clause.)
Use a condition in a query?	1. SELECT 2. CONDITIONS (Use a WHERE clause.)
Use a subquery?	SUBQUERIES
Use a wildcard?	1. SELECT 2. CONDITIONS (Use LIKE and a wildcard in a WHERE clause.)
Use an alias?	ALIASES (Enter an alias after the name of each table in the FROM clause.)
Use set operations (union, intersection, difference)?	UNION, INTERSECT, AND MINUS (Connect two SELECT commands with UNION, INTERSECT, or MINUS.)

Answers to Odd-Numbered Exercises

■ Chapter 1—Premiere Products

1. Ann Samuels, Al Williams, Sally Adams, and Mary Nelson

3. BT04, Gas Grill, $1649.89; and BZ66, Washer, $20,799.48

5. Four

7. 12489, 9/02/02, 124, Sally Adams; 12491, 9/02/02, 311, Don Charles; 12494, 9/04/02, 315, Tom Daniels; 12495, 9/04/02, 256, Ann Samuels; 12498, 9/05/02, 522, Mary Nelson; 12500, 9/05/02, 124, Sally Adams; 12504, 9/05/02, 522, Mary Nelson

9. 03, Mary Jones; 12, Miguel Diaz; 06; William Smith

■ Chapter 1—Henry Books

1. Arcade Publishing, Bantam Books, McPherson and Co., Pocket Books, Random House, Rizzoli, Schoken Books, Signet, Thames and Hudson, W.W. Norton and Co.

3. 0378, Dunwich Horror and Others; 1351, Cujo; 7443, Carrie

5. 0189, Kane and Abel; 0378, Dunwich Horror and Others; 079X, Smokescreen; 0808, Knockdown; 1351, Cujo; 1382, Marcel Duchamp; 2281, Prints of the 20th Century; 2766, Prodigal Daughter; 3350, Higher Creativity; 3743, First Among Equals; 6128, Evil Under the Sun; 7443, Carrie; 7559, Risk

7. 0180, Shyness, $6.89; 0189, Kane and Abel, $5.00; 0200, Stranger, $7.88; 0378, Dunwich Horror and Others, $17.78; 079X, Smokescreen, $4.10; etc. To calculate the remaining prices, multiply the current price in the BOOK_PRICE field by .90.

9. 0189, Kane and Abel; 0200, Stranger; 079X, Smokescreen; 0808, Knockdown; 1382, Marcel Duchamp; 138X, Death on the Nile; 2281, Prints of the 20th Century; 2766, Prodigal Daughter; 3743, First Among Equals; 6128, Evil Under the Sun; 7559, Risk; 8092, Magritte; 8720, Castle; 9611, Amerika

11. $11.05

13. 1, 0180, Shyness, 2; 2, 0189, Kane and Abel, 2; 1, 0200, Stranger, 1; 2, 0200, Stranger, 3; 2, 079X, Smokescreen, 1; 3, 079X, Smokescreen, 2; 4, 079X, Smokescreen, 3; 1, 1351, Cujo, 1; 2, 1351, Cujo, 4; 3, 1351, Cujo, 2; 2, 138X, Death on the Nile, 3; 1, 2226, Ghost from the Grand Banks, 3; 3, 2226, Ghost from the Grand Banks, 2; 4, 2226, Ghost from the Grand Banks, 1; 4, 2281, Prints of the 20th Century, 3; 3, 2766, Prodigal Daughter, 2; 1, 2908, Hymns in the Night, 3; 4, 2908, Hymns in the Night, 1; 1, 3350, Higher Creativity, 2; 2, 3906, Vortex, 1; 3, 3906, Vortex, 2; 1, 5163, Organ, 1; 4, 5790, Database Systems, 2; 2, 6128, Evil Under the Sun, 4; 3, 6128, Evil Under the Sun, 3; 2, 6328, Vixen 07, 2; 1, 669X, A Guide

to SQL, 1; 2, 6908, DOS Essentials 2; 3, 7405, Night Probe, 2; 2, 7559, Risk, 2; 2, 7947, dBASE Programming, 2; 3, 8092, Magritte, 1; 1, 8720, Castle, 3; 1, 9611, Amerika, 2

■ Chapter 2—Premiere Products

Use SQL to input the specified data.

■ Chapter 2—Henry Books

Use SQL to input the specified data.

■ Chapter 3—Premiere Products

1. SELECT PART_NUMBER, PART_DESCRIPTION
   FROM PART;
   AX12 Iron
   AZ52 Dartboard
   BA74 Basketball
   BH22 Cornpopper
   BT04 Gas Grill
   BZ66 Washer
   CA14 Griddle
   CB03 Bike
   CX11 Blender
   CZ81 Treadmill

3. SELECT LAST, FIRST
   FROM CUSTOMER
   WHERE CREDIT_LIMIT >= 800;
   Adams   Sally
   Samuels   Ann
   Charles   Don
   Williams   Al
   Adams   Sally
   Nelson   Mary
   Galvez   Mara
   Martin   Dan

5. SELECT CUSTOMER_NUMBER, LAST, FIRST
   FROM CUSTOMER
   WHERE SLSREP_NUMBER = '03' OR SLSREP_NUMBER = '12';

124	Adams	Sally
311	Charles	Don
405	Williams	Al
412	Adams	Sally
522	Nelson	Mary
622	Martin	Dan

7. SELECT PART_NUMBER, PART_DESCRIPTION
   FROM PART
   WHERE UNITS_ON_HAND >100 AND UNITS_ON_HAND < 200;

   AX12 Iron
   CX11 Blender

   SELECT PART_NUMBER, PART_DESCRIPTION
   FROM PART
   WHERE UNITS_ON_HAND BETWEEN 100 AND 200;

   AX12 Iron
   CX11 Blender

9. SELECT PART_NUMBER, PART_DESCRIPTION, (UNITS_ON_HAND *
   UNIT_PRICE)
   FROM PART
   WHERE (UNITS_ON_HAND * UNIT_PRICE) >= 1000;

AX12	Iron	2594.8
BA74	Basketball	1198
BH22	Cornpopper	2370.25
BT04	Gas Grill	1649.89
BZ66	Washer	20799.48
CA14	Griddle	3119.22
CB03	Bike	13199.56
CX11	Blender	2570.4
CZ81	Treadmill	23796.6

11. SELECT CUSTOMER_NUMBER, LAST, FIRST
    FROM CUSTOMER
    WHERE FIRST LIKE 'D%';

311	Charles	Don
622	Martin	Dan

13. SELECT *
    FROM PART
    ORDER BY ITEM_CLASS, PART_NUMBER;

BT04 Gas Grill	11	AP	2	149.99
BZ66 Washer	52	AP	3	399.99
AX12 Iron	104	HW	3	24.95
BH22 Cornpopper	95	HW	3	24.95
CA14 Griddle	78	HW	3	39.99
CX11 Blender	112	HW	3	22.95
AZ52 Dartboard	20	SG	2	12.95
BA74 Basketball	40	SG	1	29.95
CB03 Bike	44	SG	1	299.99
CZ81 Treadmill	68	SG	2	349.95

15. SELECT SUM(BALANCE)
    FROM CUSTOMER
    WHERE SLSREP_NUMBER = '12' AND BALANCE < CREDIT_LIMIT;
    1327.25

17. SELECT MIN(UNIT_PRICE)
    FROM PART;
    12.95

19. SELECT COUNT(*)
    FROM CUSTOMER;
    10

21. SELECT ITEM_CLASS, SUM(UNITS_ON_HAND * UNIT_PRICE)
    FROM PART
    GROUP BY ITEM_CLASS;

AP	22449.37
HW	10654.67
SG	38453.16

■ Chapter 3—Henry Books

1. SELECT BOOK_CODE, BOOK_TITLE
   FROM BOOK;

0180	Shyness
0189	Kane and Abel
0200	Stranger

0378 Dunwich Horror and Others
079X Smokescreen
0808 Knockdown
1351 Cujo
1382 Marcel Duchamp
138X Death on the Nile
2226 Ghost from the Grand Banks
2281 Prints of the 20th Century
2766 Prodigal Daughter
2908 Hymns to the Night
3350 Higher Creativity
3743 First Among Equals
3906 Vortex
5163 Organ
5790 Database Systems
6128 Evil Under the Sun
6328 Vixen 07
669X A Guide to SQL
6908 DOS Essentials
7405 Night Probe
7443 Carrie
7559 Risk
7947 dBASE Programming
8092 Magritte
8720 Castle
9611 Amerika

3. SELECT PUBLISHER_NAME
   FROM PUBLISHER
   WHERE PUBLISHER_STATE = 'NY';
   Arcade Publishing
   Bantam Books
   McPherson and Co.
   Pocket Books
   Random House
   Rizzoli
   Schoken Books
   Signet
   Thames and Hudson
   W.W. Norton and Co.

5. SELECT BRANCH_NAME
   FROM BRANCH
   WHERE NUMBER_EMPLOYEES >= 10;
   Henrys Downtown
   Henrys Brentwood

7. SELECT BOOK_CODE, BOOK_TITLE
   FROM BOOK
   WHERE BOOK_TYPE = 'HOR' AND PAPERBACK = 'Y';
   1351 Cujo
   7443 Carrie

9. SELECT BOOK_CODE, BOOK_TITLE, BOOK_PRICE
   FROM BOOK
   WHERE BOOK_PRICE BETWEEN 10 AND 20;
   (**Note:** The WHERE clause could also be WHERE BOOK_PRICE > 10
   AND BOOK_PRICE < 20;)

0378 Dunwich Horror and Others	19.75
1382 Marcel Duchamp	11.25
2226 Ghost from the Grand Banks	19.95
2281 Prints of the 20th Century	13.25
5163 Organ	16.95
8720 Castle	12.15
9611 Amerika	10.95

11. SELECT BOOK_CODE, BOOK_TITLE, (BOOK_PRICE * .85)
    FROM BOOK;

0180 Shyness	6.5025
0189 Kane and Abel	4.7175
0200 Stranger	7.4375
0378 Dunwich Horror and Others	16.7875
079X Smokescreen	3.8675
0808 Knockdown	4.0375
1351 Cujo	5.6525
1382 Marcel Duchamp	9.5625
138X Death on the Nile	3.3575
2226 Ghost from the Grand Banks	16.9575
2281 Prints of the 20th Century	11.2625
2766 Prodigal Daughter	4.6325
2908 Hymns to the Night	5.7375
3350 Higher Creativity	8.2875

3743 First Among Equals	3.3575
3906 Vortex	4.6325
5163 Organ	14.4075
5790 Database Systems	46.7075
6128 Evil Under the Sun	3.7825
6328 Vixen 07	4.7175
669X A Guide to SQL	20.3575
6908 DOS Essentials	17.425
7405 Night Probe	4.8025
7443 Carrie	5.7375
7559 Risk	3.3575
7947 dBASE Programming	33.915
8092 Magritte	18.6575
8720 Castle	10.3275
9611 Amerika	9.3075

13. SELECT BOOK_CODE, BOOK_TITLE
    FROM BOOK
    WHERE BOOK_TYPE IN ('FIC','MYS','ART');
    0189 Kane and Abel
    079X Smokescreen
    0808 Knockdown
    1382 Marcel Duchamp
    138X Death on the Nile
    2281 Prints of the 20th Century
    2766 Prodigal Daughter
    3743 First Among Equals
    6128 Evil Under the Sun
    7559 Risk
    8092 Magritte
    8720 Castle
    9611 Amerika

15. SELECT BOOK_CODE, BOOK_TITLE
    FROM BOOK
    WHERE BOOK_TYPE IN ('FIC','MYS','ART')
    ORDER BY BOOK_CODE DESC;
    9611 Amerika
    8720 Castle
    8092 Magritte
    7559 Risk

6128 Evil Under the Sun
3743 First Among Equals
2766 Prodigal Daughter
2281 Prints of the 20th Century
138X Death on the Nile
1382 Marcel Duchamp
0808 Knockdown
079X Smokescreen
0189 Kane and Abel

17. SELECT COUNT(DISTINCT(BOOK_TYPE))
FROM BOOK;
10

19. SELECT BOOK_TYPE, AVG(BOOK_PRICE)
FROM BOOK
GROUP BY BOOK_TYPE;

ART	15.483333
CS	34.825
FIC	7.61
HOR	10.475
MUS	16.95
MYS	4.33
POE	6.75
PSY	8.7
SFI	19.95
SUS	5.55

21. SELECT BOOK_TITLE
FROM BOOK
WHERE BOOK_PRICE =
(SELECT MAX(BOOK_PRICE)
FROM BOOK);
Database Systems

23. SELECT BRANCH_NAME
FROM BRANCH
WHERE NUMBER_EMPLOYEES IN
(SELECT MIN(NUMBER_EMPLOYEES)
FROM BRANCH);
Henrys On The Hill

1. SELECT ORDER_NUMBER, ORDER_DATE,
   CUSTOMER.CUSTOMER_NUMBER, LAST, FIRST
   FROM ORDERS, CUSTOMER
   WHERE ORDERS.CUSTOMER_NUMBER =
   CUSTOMER.CUSTOMER_NUMBER;
   12489 02-SEP-02 124 Adams Sally
   12491 02-SEP-02 311 Charles Don
   12494 04-SEP-02 315 Daniels Tom
   12495 04-SEP-02 256 Samuels Ann
   12498 05-SEP-02 522 Nelson Mary
   12500 05-SEP-02 124 Adams Sally
   12504 05-SEP-02 522 Nelson Mary

3. SELECT ORDERS.ORDER_NUMBER, ORDER_DATE,
   PART.PART_NUMBER, NUMBER_ORDERED, QUOTED_PRICE
   FROM ORDERS, ORDER_LINE, PART
   WHERE ORDERS.ORDER_NUMBER =
   ORDER_LINE.ORDER_NUMBER
   AND PART.PART_NUMBER = ORDER_LINE.PART_NUMBER;

12489	02-SEP-02	AX12	11	21.95
12491	02-SEP-02	BT04	1	149.99
12491	02-SEP-02	BZ66	1	399.99
12494	04-SEP-02	CB03	4	279.99
12495	04-SEP-02	CX11	2	22.95
12498	05-SEP-02	AZ52	2	12.95
12498	05-SEP-02	BA74	4	24.95
12500	05-SEP-02	BT04	1	149.99
12504	05-SEP-02	CZ81	2	325.99

5. SELECT CUSTOMER_NUMBER, LAST, FIRST
   FROM CUSTOMER
   WHERE EXISTS
   (SELECT *
   FROM ORDERS
   WHERE ORDERS.CUSTOMER_NUMBER =
   CUSTOMER.CUSTOMER_NUMBER
   AND ORDER_DATE = '05-SEP-02');
   124 Adams   Sally
   522 Nelson   Mary

7. SELECT ORDER_LINE.ORDER_NUMBER, ORDER_DATE,
PART.PART_NUMBER, PART_DESCRIPTION, ITEM_CLASS
FROM ORDER_LINE, ORDERS, PART
WHERE ORDER_LINE.ORDER_NUMBER =
ORDERS.ORDER_NUMBER
AND ORDER_LINE.PART_NUMBER = PART.PART_NUMBER;

12489	02-SEP-02	AX12 Iron	HW
12491	02-SEP-02	BT04 Gas Grill	AP
12491	02-SEP-02	BZ66 Washer	AP
12494	04-SEP-02	CB03 Bike	SG
12495	04-SEP-02	CX11 Blender	HW
12498	05-SEP-02	AZ52 Dartboard	SG
12498	05-SEP-02	BA74 Basketball	SG
12500	05-SEP-02	BT04 Gas Grill	AP
12504	05-SEP-02	CZ81 Treadmill	SG

9. SELECT SLSREP_NUMBER, LAST, FIRST
FROM SALES_REP
WHERE SLSREP_NUMBER IN
(SELECT SLSREP_NUMBER
FROM CUSTOMER
WHERE CREDIT_LIMIT IN
(SELECT CREDIT_LIMIT
FROM CUSTOMER
WHERE CREDIT_LIMIT = 2000));

03   Jones   Mary

11. SELECT CUSTOMER.CUSTOMER_NUMBER, LAST, FIRST
FROM CUSTOMER, ORDER_LINE, ORDERS, PART
WHERE CUSTOMER.CUSTOMER_NUMBER =
ORDERS.CUSTOMER_NUMBER
AND ORDERS.ORDER_NUMBER = ORDER_LINE.ORDER_NUMBER
AND PART.PART_NUMBER = ORDER_LINE.PART_NUMBER
AND PART_DESCRIPTION = 'Iron';

124   Adams   Sally

13. SELECT PART.PART_DESCRIPTION, PART.PART_NUMBER,
    ORDERS.ORDER_NUMBER, ORDER_DATE
    FROM ORDERS, ORDER_LINE, CUSTOMER, PART
    WHERE PART.PART_NUMBER = ORDER_LINE.PART_NUMBER
    AND ORDERS.ORDER_NUMBER = ORDER_LINE.ORDER_NUMBER
    AND CUSTOMER.CUSTOMER_NUMBER =
    ORDERS. CUSTOMER_NUMBER
    AND CUSTOMER.LAST = 'Nelson'
    AND CUSTOMER.FIRST = 'Mary'
    AND PART_DESCRIPTION = 'Treadmill';
    Treadmill   CZ81 12504 05-SEP-02

15. SELECT ORDER_NUMBER, ORDER_DATE
    FROM ORDERS, CUSTOMER
    WHERE ORDERS.CUSTOMER_NUMBER =
    CUSTOMER. CUSTOMER_NUMBER
    AND LAST = 'Nelson'
    AND FIRST = 'Mary'
    UNION
    SELECT ORDERS.ORDER_NUMBER, ORDER_DATE
    FROM ORDERS, ORDER_LINE, PART
    WHERE ORDERS.ORDER_NUMBER = ORDER_LINE.ORDER_NUMBER
    AND PART.PART_NUMBER = ORDER_LINE.PART_NUMBER
    AND PART_DESCRIPTION = 'Iron';
    12489 02-SEP-02
    12498 05-SEP-02
    12504 05-SEP-02

17. SELECT ORDER_NUMBER, ORDER_DATE
    FROM ORDERS, CUSTOMER
    WHERE ORDERS.CUSTOMER_NUMBER =
    CUSTOMER. CUSTOMER_NUMBER
    AND LAST = 'Nelson'
    AND FIRST = 'Mary'
    MINUS
    SELECT ORDERS.ORDER_NUMBER, ORDER_DATE
    FROM ORDERS, ORDER_LINE, PART
    WHERE ORDERS.ORDER_NUMBER = ORDER_LINE.ORDER_NUMBER
    AND PART.PART_NUMBER = ORDER_LINE.PART_NUMBER
    AND PART_DESCRIPTION = 'Iron';
    12498   05-SEP-02
    12504   05-SEP-02

19. SELECT PART_NUMBER, PART_DESCRIPTION, UNIT_PRICE,
ITEM_CLASS
FROM PART
WHERE UNIT_PRICE > ANY
(SELECT UNIT_PRICE
FROM PART
WHERE ITEM_CLASS = 'HW');

AX12	Iron	24.95	HW
BA74	Basketball	29.95	SG
BH22	Cornpopper	24.95	HW
BT04	Gas Grill	149.99	AP
BZ66	Washer	399.99	AP
CA14	Griddle	39.99	HW
CB03	Bike	299.99	SG
CZ81	Treadmill	349.95	SG

The question asks for the list of parts that have a unit price that is greater than any part in item class HW.

■ Chapter 4—Henry Books

1. SELECT BOOK.BOOK_CODE, BOOK_TITLE,
PUBLISHER.PUBLISHER_CODE, PUBLISHER_NAME
FROM BOOK, PUBLISHER
WHERE PUBLISHER.PUBLISHER_CODE = BOOK.PUBLISHER_CODE;

0180 Shyness	BB	Bantam Books
0189 Kane and Abel	PB	Pocket Books
0200 Stranger	BB	Bantam Books
0378 Dunwich Horror and Others	PB	Pocket Books
079X Smokescreen	PB	Pocket Books
0808 Knockdown	PB	Pocket Books
1351 Cujo	SI	Signet
1382 Marcel Duchamp	PB	Pocket Books
138X Death on the Nile	BB	Bantam Books
2226 Ghost from the Grand Banks	BB	Bantam Books
2281 Prints of the 20th Century	PB	Pocket Books
2766 Prodigal Daughter	PB	Pocket Books
2908 Hymns to the Night	BB	Bantam Books
3350 Higher Creativity	PB	Pocket Books
3743 First Among Equals	PB	Pocket Books

3906 Vortex	BB	Bantam Books
5163 Organ	SI	Signet
5790 Database Systems	BF	Best and Furrow
6128 Evil Under the Sun	PB	Pocket Books
6328 Vixen 07	BB	Bantam Books
669X A Guide to SQL	BF	Best and Furrow
6908 DOS Essentials	BF	Best and Furrow
7405 Night Probe	BB	Bantam Books
7443 Carrie	SI	Signet
7559 Risk	PB	Pocket Books
7947 dBASE Programming	BF	Best and Furrow
8092 Magritte	SI	Signet
8720 Castle	BB	Bantam Books
9611 Amerika	BB	Bantam Books

3. SELECT BOOK_TITLE, BOOK.BOOK_CODE
FROM BOOK, PUBLISHER
WHERE BOOK.PUBLISHER_CODE = PUBLISHER.PUBLISHER_CODE
AND BOOK_PRICE > 10
AND PUBLISHER_NAME = 'Bantam Books';

Ghost from the Grand Banks	2226
Castle	8720
Amerika	9611

5. SELECT BOOK_TITLE
FROM BOOK, PUBLISHER
WHERE PUBLISHER.PUBLISHER_CODE = BOOK.PUBLISHER_CODE
AND BOOK_TYPE = 'CS'
AND PUBLISHER_NAME = 'Best and Furrow';

Database Systems
A Guide to SQL
DOS Essentials
dBASE Programming

7. SELECT BOOK_TITLE
FROM BOOK
WHERE EXISTS
(SELECT *
FROM WROTE
WHERE BOOK.BOOK_CODE = WROTE.BOOK_CODE
AND AUTHOR_NUMBER = '01');

Kane and Abel
Prodigal Daughter
First Among Equals

9. SELECT F.BOOK_CODE, S.BOOK_CODE
FROM INVENT F, INVENT S
WHERE F.BRANCH_NUMBER = S.BRANCH_NUMBER
AND F.BOOK_CODE < S.BOOK_CODE;

The query will select 138 rows. The first 10 rows are:

0180	0200
0189	0200
0189	079X
0200	079X
0180	1351
0200	1351
0189	1351
0200	1351
079X	1351
079X	1351

11. SELECT UNITS_ON_HAND, BOOK_TITLE, AUTHOR_LAST
FROM INVENT, BOOK, WROTE, AUTHOR
WHERE INVENT.BOOK_CODE = BOOK.BOOK_CODE
AND BOOK.BOOK_CODE = WROTE.BOOK_CODE
AND WROTE.AUTHOR_NUMBER = AUTHOR.AUTHOR_NUMBER
AND BRANCH_NUMBER = '4'
AND PAPERBACK = 'Y';

3	Smokescreen	Francis
3	Prints of the 20th Century	Castleman
1	Hymns to the Night	Novalis

13. SELECT BOOK.BOOK_CODE, BOOK_TITLE
FROM BOOK
WHERE BOOK_PRICE > 5
INTERSECT
SELECT BOOK_CODE, BOOK_TITLE
FROM BOOK, PUBLISHER
WHERE BOOK.PUBLISHER_CODE = PUBLISHER.PUBLISHER_CODE
AND PUBLISHER CITY = 'New York';

0180 Shyness
0189 Kane and Abel
0200 Stranger
0378 Dunwich Horror and Others
1351 Cujo
1382 Marcel Duchamp

2226 Ghost from the Grand Banks
2281 Prints of the 20th Century
2766 Prodigal Daughter
2908 Hymns to the Night
3350 Higher Creativity
3906 Vortex
5163 Organ
6328 Vixen 07
7405 Night Probe
7443 Carrie
8092 Magritte
8720 Castle
9611 Amerika

15. SELECT BOOK_TITLE, PUBLISHER_CODE
FROM BOOK
WHERE BOOK_PRICE > ALL
(SELECT BOOK_PRICE
FROM BOOK
WHERE BOOK_TYPE = 'HOR');

Ghost from the Grand Banks	BB
Database Systems	BF
A Guide to SQL	BF
DOS Essentials	BF
dBASE Programming	BF
Magritte	SI

■ Chapter 5—Premiere Products

1. UPDATE PART
SET PART_DESCRIPTION = 'Oven'
WHERE PART_NUMBER = 'BT04';
1 row updated.

SELECT *
FROM PART;

AX12	Iron	104	HW	3	24.95
AZ52	Dartboard	20	SG	2	12.95
BA74	Basketball	40	SG	1	29.95
BH22	Cornpopper	95	HW	3	24.95
BT04	Oven	11	AP	2	149.99
BZ66	Washer	52	AP	3	399.99

CA14	Griddle	78	HW	3	39.99
CB03	Bike	44	SG	1	299.99
CX11	Blender	112	HW	3	22.95
CZ81	Treadmill	68	SG	2	349.95

3. INSERT INTO ORDERS
VALUES
('12600','06-SEP-2002','311');
1 row created.

INSERT INTO ORDER_LINE
VALUES
('12600','AX12',5,13.95);
1 row created.

INSERT INTO ORDER_LINE
VALUES
('12600','BA74',3,4.50);
1 row created.

SELECT *
FROM ORDER_LINE;

12489	AX12	11	21.95
12491	BT04	1	149.99
12491	BZ66	1	399.99
12494	CB03	4	279.99
12495	CX11	2	22.95
12498	AZ52	2	12.95
12498	BA74	4	24.95
12500	BT04	1	149.99
12504	CZ81	2	325.99
12600	AX12	5	13.95
12600	BA74	3	4.5

SELECT *
FROM ORDERS;

12489	02-SEP-02	124
12491	02-SEP-02	311
12494	04-SEP-02	315
12495	04-SEP-02	256
12498	05-SEP-02	522
12500	05-SEP-02	124
12504	05-SEP-02	522
12600	06-SEP-02	311

5. CREATE TABLE SPGOOD
(PART_NUMBER CHAR(4),
PART_DESCRIPTION CHAR(12),
UNIT_PRICE NUMBER(6,2));
Table created.

INSERT INTO SPGOOD
SELECT PART_NUMBER, PART_DESCRIPTION, UNIT_PRICE
FROM PART
WHERE ITEM_CLASS = 'SG';
4 rows created.

SELECT *
FROM SPGOOD;

AZ52 Dartboard	12.95
BA74 Basketball	29.95
CB03 Bike	299.99
CZ81 Treadmill	349.95

7. ALTER TABLE PART
ADD ALLOCATION NUMBER(3,0);
Table altered.

UPDATE PART
SET ALLOCATION = 0;
10 rows updated.

SELECT SUM(NUMBER_ORDERED)
FROM ORDER_LINE
WHERE PART_NUMBER = 'BT04';
2

UPDATE PART
SET ALLOCATION = 2
WHERE PART_NUMBER = 'BT04';
1 row updated.

SELECT *
FROM PART;

AX12	Iron	104	HW	3	24.95	0
AZ52	Dartboard	20	SG	2	12.95	0
BA74	Basketball	40	SG	1	29.95	0

BH22	Cornpopper	95	HW	3	24.95	0
BT04	Gas Grill	11	AP	2	149.99	2
BZ66	Washer	52	AP	3	399.99	0
CA14	Griddle	78	HW	3	39.99	0
CB03	Bike	44	SG	1	299.99	0
CX11	Blender	112	HW	3	22.95	0
CZ81	Treadmill	68	SG	2	349.95	0

■ Chapter 5—Henry Books

1. UPDATE INVENT
   SET UNITS_ON_HAND = 5
   WHERE BRANCH_NUMBER = '1';
   10 rows updated.

   SELECT *
   FROM INVENT
   WHERE BRANCH_NUMBER = '1';

0180	1	5
0200	1	5
1351	1	5
2226	1	5
2908	1	5
3350	1	5
5163	1	5
669X	1	5
8720	1	5
9611	1	5

3. INSERT INTO BOOK
   VALUES
   ('9700','Using Microsoft Access 2000','BF','CS',19.97,'Y');
   1 row created.

   INSERT INTO INVENT
   VALUES
   ('9700','1',4);
   1 row created.

   INSERT INTO WROTE
   VALUES
   ('9700','07',2);
   1 row created.

5. CREATE TABLE FICTION
    (BOOK_CODE CHAR(4),
    BOOK_TITLE CHAR(30),
    BOOK_PRICE NUMBER(4,2));
    Table created.

    INSERT INTO FICTION
    SELECT BOOK_CODE, BOOK_TITLE, BOOK_PRICE
    FROM BOOK
    WHERE BOOK_TYPE = 'FIC';
    5 rows created.

    SELECT *
    FROM FICTION;

0189	Kane and Abel	5.55
2766	Prodigal Daughter	5.45
3743	First Among Equals	3.95
8720	Castle	12.15
9611	Amerika	10.95

7. UPDATE FICTION
    SET BOOK_PRICE = NULL
    WHERE BOOK_TITLE = 'Amerika';
    1 row updated.

    SELECT *
    FROM FICTION;

0189	Kane and Abel	6.22
2766	Prodigal Daughter	6.1
3743	First Among Equals	4.42
8720	Castle	13.61
9611	Amerika	

9. UPDATE FICTION
    SET BEST_SELLER = 'Y'
    WHERE BOOK_TITLE = 'Kane and Abel';
    1 row updated.

    SELECT *
    FROM FICTION;

0189	Kane and Abel	6.22	Y
2766	Prodigal Daughter	6.1	N
3743	First Among Equals	4.42	N
8720	Castle	13.61	N
9611	Amerika		N

11. ALTER TABLE FICTION
MODIFY BEST_SELLER NOT NULL;
Table altered.

13. DROP TABLE FICTION;
Table dropped.

■ Chapter 6

Premiere Products

1a. CREATE VIEW SMALLCST AS
SELECT CUSTOMER_NUMBER, LAST, FIRST, STREET, BALANCE,
CREDIT_LIMIT
FROM CUSTOMER
WHERE CREDIT_LIMIT <= 1000;
View created.

1b. SELECT CUSTOMER_NUMBER, LAST, FIRST
FROM SMALLCST
WHERE BALANCE > CREDIT_LIMIT;
315 Daniels Tom
622 Martin Dan

1c. SELECT CUSTOMER_NUMBER, LAST, FIRST
FROM CUSTOMER
WHERE CREDIT_LIMIT <= 1000
AND BALANCE > CREDIT_LIMIT;
315 Daniels Tom
622 Martin Dan

1d. There are no problems updating the data in the SMALLCST view because the
view includes the primary key of the CUSTOMER table as one of its fields. The
only potential problem is that a customer can be added to the base table with a
credit limit of more than $1,000, in which case that customer's row disappears
from the view.

3a. CREATE VIEW ORDTOT (ORDER_NUMBER, ORDER_TOTAL) AS
SELECT ORDER_NUMBER, SUM(NUMBER_ORDERED *
QUOTED_PRICE)
FROM ORDER_LINE
GROUP BY ORDER_NUMBER;
View created.

3b. SELECT ORDER_NUMBER, ORDER_TOTAL
FROM ORDTOT
WHERE ORDER_TOTAL > 500
ORDER BY ORDER_NUMBER;
12491   549.98
12494   1119.96
12504   651.98

3c. SELECT ORDER_NUMBER, SUM(NUMBER_ORDERED *
QUOTED_PRICE) AS ORDER_TOTAL
FROM ORDER_LINE
GROUP BY ORDER_NUMBER
HAVING SUM(NUMBER_ORDERED * QUOTED_PRICE) > 500;
12491   549.98
12494   1119.96
12504   651.98

3d. You cannot update data in the ORDER_LINE table using the ORDTOT view
because the view does not contain the primary key of the base table and the view
includes a computed column.

5. REVOKE SELECT ON PART FROM STILLWELL;

7. DROP INDEX PARTIND3;

9. ALTER TABLE CUSTOMER
ADD CHECK (CREDIT_LIMIT IN (750, 1000, 1500, 2000))
PRIMARY KEY (CUSTOMER_NUMBER)
FOREIGN KEY (SLSREP_NUMBER) REFERENCES SALES_REP;

■ Chapter 6—Henry Books

1a. CREATE VIEW BANTAM AS
SELECT BOOK_CODE, BOOK_TITLE, BOOK_TYPE, BOOK_PRICE
FROM BOOK
WHERE PUBLISHER_CODE = 'BB';
View created.

1b. SELECT BOOK_CODE, BOOK_TYPE, BOOK_TITLE, BOOK_PRICE
FROM BANTAM
WHERE BOOK_PRICE < 10;
0180   Shyness   7.65
0200   Stranger   8.75

138X	Death on the Nile	3.95
2908	Hymns to the Night	6.75
3906	Vortex	5.45
6328	Vixen 07	5.55
7405	Night Probe	5.65

1c. SELECT BOOK_CODE, BOOK_TITLE, BOOK_PRICE
FROM BOOK
WHERE PUBLISHER_CODE = 'BB'
AND BOOK_PRICE < 10;

0180	Shyness	7.65
0200	Stranger	8.75
138X	Death on the Nile	3.95
2908	Hymns to the Night	6.75
3906	Vortex	5.45
6328	Vixen 07	5.55
7405	Night Probe	5.65

1d. You can use the BANTAM view to update data in the BOOK table because the view contains the primary key of the underlying base table.

3a. CREATE VIEW VALUE (BRANCH_NUMBER, TOTAL_COUNT) AS
SELECT BRANCH_NUMBER, SUM(UNITS_ON_HAND)
FROM INVENT
GROUP BY BRANCH_NUMBER;
View created.

3b. SELECT *
FROM VALUE;

1	19
2	26
3	16
4	10

3c. SELECT BRANCH_NUMBER, SUM(UNITS_ON_HAND)
FROM INVENT
GROUP BY BRANCH_NUMBER;

1	19
2	26
3	16
4	10

3d. You cannot use the VALUE view to update data in the INVENT table because the view contains a computed column.

5. REVOKE INDEX ON BOOK FROM VERNER;
   REVOKE ALTER ON AUTHOR FROM VERNER;

7. DROP INDEX BOOKIND3;

9a. ALTER TABLE BOOK
    ADD CHECK (BOOK_CODE <= '9999');

9b. ALTER TABLE BOOK
    ADD CHECK (BOOK_TYPE IN
    ('PSY', 'FIC', 'HOR', 'MYS', 'ART', 'POE', 'SUS', 'MUS', 'CS'));

9c. ALTER TABLE BOOK
    ADD CHECK (PAPERBACK IN ('Y', 'N'));

9d. ALTER TABLE BRANCH
    ADD CHECK (BRANCH_NUMBER IN ('1', '2', '3', '4'));

9e. ALTER TABLE WROTE
    ADD CHECK (SEQUENCE_NUMBER IN (1, 2));

■ Chapter 7—Premiere Products

1. SELECT RTRIM(FIRST)||''||RTRIM(LAST), STREET,
   RTRIM(CITY)||','||RTRIM(STATE)||''||RTRIM(ZIP_CODE)
   FROM CUSTOMER;

Sally Adams	481 Oak	Lansing, MI 49224
Ann Samuels	215 Pete	Grant, MI 49219
Don Charles	48 College	Ira, MI 49034
Tom Daniels	914 Cherry	Kent, MI 48391
Al Williams	519 Watson	Grant, MI 49219
Sally Adams	16 Elm	Lansing, MI 49224
Mary Nelson	108 Pine	Ada, MI 49441
Tran Dinh	808 Ridge	Harper, MI 48421
Mara Galvez	512 Pine	Ada, MI 49441
Dan Martin	419 Chip	Grant, MI 49219

3. COLUMN CUSTOMER_NAME HEADING 'CUSTOMER|NAME'
   COLUMN STREET_ADDRESS HEADING 'CUSTOMER|ADDRESS'
   COLUMN CITY_STATE_ZIP HEADING 'CUSTOMER|CITY/STATE/ZIP'
   SELECT *
   FROM REPORT2;

CUSTOMER NAME	CUSTOMER ADDRESS	CUSTOMER CITY/STATE/ZIP
Sally Adams	481 Oak	Lansing, MI 49224
Ann Samuels	215 Pete	Grant, MI 49219
Don Charles	48 College	Ira, MI 49034
Tom Daniels	914 Cherry	Kent, MI 48391
Al Williams	519 Watson	Grant, MI 49219
Sally Adams	16 Elm	Lansing, MI 49224
Mary Nelson	108 Pine	Ada, MI 49441
Tran Dinh	808 Ridge	Harper, MI 48421
Mara Galvez	512 Pine	Ada, MI 49441
Dan Martin	419 Chip	Grant, MI 49219

5. CREATE VIEW REPORT3 (CUSTOMER_NAME, CREDIT_LIMIT, BALANCE) AS
SELECT RTRIM(FIRST)||''||RTRIM(LAST), CREDIT_LIMIT, BALANCE
FROM CUSTOMER;
View created.

COLUMN CUSTOMER_NAME HEADING 'CUSTOMER|NAME'
COLUMN CREDIT_LIMIT HEADING 'CUSTOMER|CREDIT LIMIT'
    FORMAT $9,990.99
COLUMN BALANCE HEADING 'CUSTOMER|BALANCE'
    FORMAT $9,990.99

SELECT *
FROM REPORT3;

CUSTOMER NAME	CUSTOMER CREDIT LIMIT	CUSTOMER BALANCE
Sally Adams	$1,000.00	$818.75
Ann Samuels	$1,500.00	$21.50
Don Charles	$1,000.00	$825.75
Tom Daniels	$750.00	$770.75
Al Williams	$1,500.00	$402.75
Sally Adams	$2,000.00	$1,817.50
Mary Nelson	$1,500.00	$98.75
Tran Dinh	$750.00	$402.40
Mara Galvez	$1,000.00	$114.60
Dan Martin	$1,000.00	$1,045.75

10 rows selected.

7. SET FEEDBACK OFF
SELECT *
FROM REPORT3;

Thu Jul 27   page 1

CUSTOMER CREDIT LIMITS
AND BALANCES

CUSTOMER NAME	CUSTOMER CREDIT LIMIT	CUSTOMER BALANCE
Sally Adams	$1000.00	$818.75
Ann Samuels	$1500.00	$21.50
Don Charles	$1000.00	$825.75
Tom Daniels	$750.00	$770.75
Al Williams	$1500.00	$402.75
Sally Adams	$2000.00	$1,817.50
Mary Nelson	$1500.00	$98.75
Tran Dinh	$750.00	$402.40
Mara Galvez	$1000.00	$114.60
Dan Martin	$1000.00	$1,045.75

■ Chapter 7—Henry Books

1. The view to create the report is:

```
CREATE VIEW BRANCH_REPORT (BRANCH_NUMBER,
BOOK_TITLE, PUBLISHER_NAME, PUBLISHER_LOCATION,
PRICE, UNITS_ON_HAND) AS
SELECT BRANCH.BRANCH_NUMBER, BOOK_TITLE,
PUBLISHER_NAME, RTRIM(PUBLISHER_CITY)||',
'||RTRIM(PUBLISHER_STATE), BOOK_PRICE, UNITS_ON_HAND
FROM BRANCH, INVENT, BOOK, PUBLISHER
WHERE BRANCH.BRANCH_NUMBER = INVENT.BRANCH_NUMBER
AND INVENT.BOOK_CODE = BOOK.BOOK_CODE
AND BOOK. PUBLISHER_CODE = PUBLISHER.PUBLISHER_CODE;
```

View created.

The script to format the report is:
```
CLEAR COLUMNS
CLEAR BREAK
CLEAR COMPUTE
TTITLE OFF
```

```
SELECT *
FROM BRANCH_REPORT
ORDER BY BRANCH_NUMBER

COLUMN BRANCH_NUMBER HEADING 'BRANCH|NUMBER'
FORMAT A6
COLUMN BOOK_TITLE HEADING 'BOOK TITLE' FORMAT A15
COLUMN PUBLISHER_NAME HEADING 'PUBLISHER NAME'
FORMAT A15
COLUMN PUBLISHER_LOCATION HEADING
'PUBLISHER|LOCATION' FORMAT A12
COLUMN PRICE HEADING 'PRICE' FORMAT $990.99
COLUMN UNITS_ON_HAND HEADING 'UNITS|ON HAND'
FORMAT 99

SET LINESIZE 70
SET PAGESIZE 66
SET PAUSE OFF

TTITLE 'INVENTORY LIST|HENRY BOOKS'

BREAK ON REPORT ON BRANCH_NUMBER SKIP 1

COMPUTE SUM OF UNITS_ON_HAND ON BRANCH_NUMBER
COMPUTE SUM OF UNITS_ON_HAND ON REPORT
```

■ Chapter 8—Premiere Products

1a.
```
EXEC SQL
      SELECT PART_DESCRIPTION, UNIT_PRICE
            INTO :W-PART-DESCRIPTION, :W-UNIT-PRICE
            FROM PART
            WHERE PART_NUMBER = :W-PART-NUMBER
END-EXEC.
```

1b.
```
EXEC SQL
SELECT ORDER_DATE, ORDERS.CUSTOMER_NUMBER, LAST,
            FIRST
      INTO :W-ORDER-DATE, W-CUSTOMER-NUMBER, W-LAST,
            W-FIRST
      FROM ORDERS, CUSTOMER
      WHERE ORDER_NUMBER = :W-ORDER-NUMBER
      AND ORDERS.CUSTOMER_NUMBER =
            CUSTOMER.CUSTOMER_NUMBER
END-EXEC.
```

```
1c.  EXEC SQL
         INSERT
         INTO PART
              VALUES (:W-PART-NUMBER, :W-PART-DESCRIPTION,
                   :W-UNITS-ON-HAND, :W-ITEM-CLASS,
                   :W-WAREHOUSE-NUMBER, :W-UNIT-PRICE)
     END-EXEC.

1d.
     EXEC SQL
         UPDATE PART
              SET PART_DESCRIPTION = :W-PART-DESCRIPTION
              WHERE PART_NUMBER = :W-PART-NUMBER
     END-EXEC.

1e.
     EXEC SQL
         UPDATE PART
              SET UNIT_PRICE = UNIT_PRICE * 1.05
              WHERE ITEM_CLASS = 'HW'
     END-EXEC.

1f.
     EXEC SQL
         DELETE
              FROM PART
              WHERE PART_NUMBER = :W-PART-NUMBER
     END-EXEC.
```

■ Chapter 8—Henry Books

```
1a.  EXEC SQL
         SELECT PUBLISHER_NAME, PUBLISHER_CITY
              INTO :W-PUBLISHER-NAME, :W-PUBLISHER-CITY
              FROM PUBLISHER
              WHERE PUBLISHER_CODE = :W-PUBLISHER-CODE
     END-EXEC.
```

1b. EXEC SQL
     SELECT BOOK_TITLE, BOOK.PUBLISHER_CODE, PUBLISHER_NAME
         INTO :W-BOOK-TITLE, W-PUBLISHER-CODE,
            W-PUBLISHER-NAME
         FROM BOOK, PUBLISHER
         WHERE BOOK_CODE = :W-BOOK-CODE
         AND BOOK.BOOK_CODE = PUBLISHER.BOOK_CODE
     END-EXEC.

1c. EXEC SQL
         INSERT
         INTO BRANCH
            VALUES (:W-BRANCH-NUMBER, :W-BRANCH-NAME,
               :W-BRANCH-LOCATION,
               :W-NUMBER-EMPLOYEES)
     END-EXEC.

1d.
     EXEC SQL
         UPDATE BOOK
            SET BOOK_TITLE = :W-BOOK-TITLE
            WHERE BOOK_CODE = :W-BOOK-CODE
     END-EXEC.

1e.
     EXEC SQL
         UPDATE BOOK
            SET BOOK_PRICE = BOOK_PRICE * 1.03
            WHERE PUBLISHER_CODE = 'BB'
     END-EXEC.

1f.
     EXEC SQL
         DELETE
            FROM BOOK
            WHERE BOOK_CODE = :W-BOOK-CODE
     END-EXEC.

INDEX

greater than or equal to (@gt=)
operator, 204
greater than or equal to (@lt=)
operator, 47
GROUP BY clause, 68, 82
GROUP BY command, 63–64
grouping records, 63–67
groups
individual totals, 64
limiting, 65–67
performing calculations on entire, 63

HAVING clause, 65–67, 68, 82
heading, 3
HEADING clause, 157
HEADING command, 169
Henry Books database, 8–11
host variables, 176
OF command, 182
naming, 176
qualification, 182
HOUSEWARES table, 119
HOUSEWARES view, 119–122,
124–126, 208

IN clause, 53–54
IN conditions, 205
IN operator, 68, 77–78
INCLUDE statement, 193
INCLUDEDECSALES_REP
statement, 180
INDEX privilege, 131
indexes
advantages and disadvantages, 136
columns, 135
creation of, 136–137
creation privilege, 131
data retrieval, 136
deleting, 137, 209
disk space, 136
unique, 137–138

unique numbers, 134
updating, 136
INPUT command, 25
INSERT command, 28–30, 33–36,
103–104, 106–107, 131, 144, 183
nulls, 31
SELECT clause, 114
INSERT INTO command, 29, 211
inserting line in command, 25
INTEGER data type, 27, 208
integrity constraint, 141–145
integrity support, 141
INTERSECT operator, 88, 91, 215–216
intersection, 88, 91
INTO clause, 176, 189, 193
IS NOT NULL operator, 67, 68
IS NULL operator, 68
IS-DATA-VALID flag, 185
ITEM_CLASS column, 56, 124–125

joining tables
conditions in WHERE clause, 77–78
qualifying columns, 74–75
restricting rows, 76
views, 122
joins
base tables, 126–128
four tables, 86–88
retrieving single row from, 182

keys, sorting on multiple, 55

L command, 25–26, 29
LAST column, 23, 31, 203
legal values constraint, 142
less than (@lt) operator, 47, 204
less than or equal to (@lt=) operator, 47, 204
library, 175
LIKE conditions, 205
LIKE operator, 52–53, 68